Computing for Biologists

Python Programming and Principles

Computing is revolutionizing the practice of biology. This book, which assumes no prior computing experience, provides students with the tools to write their own Python programs and to understand fundamental concepts in computational biology and bioinformatics.

Each major part of the book begins with a compelling biological question, followed by the algorithmic ideas and programming tools necessary to explore it: the origins of pathogenicity are examined using gene finding, the evolutionary history of sex determination systems is studied using sequence alignment, and the origin of modern humans is addressed using phylogenetic methods. In addition to providing general programming skills, this book explores the design of efficient algorithms, simulation, NP-hardness, and the maximum likelihood method, among other key concepts and methods.

Easy-to-read and designed to equip students with the skills to write programs for solving a range of biological problems, the book is accompanied by numerous programming exercises, available at www.cs.hmc.edu/CFB.

Ran Libeskind-Hadas is the R. Michael Shanahan Professor of Computer Science at Harvey Mudd College, USA, working in the areas of algorithms and computational biology. He is a recipient of both the Iris and Howard Critchell Professorship and the Joseph B. Platt Professorship for teaching, as well as the Distinguished Alumni Educator Award from the University of Illinois Urbana-Champaign Department of Computer Science.

Eliot Bush is Associate Professor of Biology at Harvey Mudd College, USA. His main research interest is the study of evolution. Among other things he has modeled the evolution of metabolism, characterized DNA methylation patterns in insects, developed algorithms for studying substitution bias in DNA, and analyzed a 30 million-year-old primate fossil. His teaching interests focus on incorporating computers and programming assignments into biology coursework.

Computing

for Biologists

Python Programming and Principles

Ran Libeskind-Hadas

Department of Computer Science,
Harvey Mudd College

Eliot Bush

Department of Biology,
Harvey Mudd College

CAMBRIDGE
UNIVERSITY PRESS

CAMBRIDGE
UNIVERSITY PRESS

University Printing House, Cambridge CB2 8BS, United Kingdom

Cambridge University Press is part of the University of Cambridge.

It furthers the University's mission by disseminating knowledge in the pursuit of
education, learning and research at the highest international levels of excellence.

www.cambridge.org
Information on this title: www.cambridge.org/9781107042827

First published 2014

Printed in the United States of America by Sheridan Books, Inc.

A catalogue record for this publication is available from the British Library

Library of Congress Cataloguing in Publication data
Libeskind-Hadas, Ran.
Computing for biologists / Ran Libeskind-Hadas, Department of Computer Science, Harvey Mudd College,
Eliot Bush, Department of Biology, Harvey Mudd College.
    pages   cm
Includes index.
ISBN 978-1-107-04282-7 (Hardback) – ISBN 978-1-107-64218-8 (Paperback)
1. Biology–Data processing.   2. Python (Computer program language)   3. Computer programming.
I. Bush, Eliot Christen.   II. Title.
QH324.2.L53 2014
570.285–dc23   2014014322

ISBN 978-1-107-04282-7 Hardback
ISBN 978-1-107-64218-8 Paperback

Additional resources for this publication at www.cambridge.org/c4b

CONTENTS

PREFACE
What's this book about?

The computer is the most powerful general-purpose tool available to biologists.

In part, this is due to the continuing rapid growth of biological data. For example, at the time of writing, the GenBank database had over 100 *million* genetic sequences with over 100 *billion* DNA characters. Among the contents of that database are genes from many organisms, annotated with what's known about their function.

Imagine that you're studying a bacterium and wish to understand what causes it to be infectious. One promising approach is to identify genes in the bacterium and compare these to known genes in GenBank. If you're able to find similar genes whose function is known, it will tell you a great deal about the role of the genes in your bacterium. This approach represents a computational challenge, and is, in fact, the topic of Part I of this book.

But searching enormous databases is not the only reason that computers are so useful to biologists. Many biological problems have a large number of different possible solutions and only a computer – programmed with carefully designed computational recipes or "algorithms" – has any chance of finding the right one. For example, biological molecules such as proteins and RNA fold into complex shapes that strongly impact their function. Computational techniques have been developed to predict how these molecules fold. Such techniques help us understand how proteins and RNA work and can even help us design new molecules to treat disease.

Simply put, computing is revolutionizing the practice of biology.

In order to fully appreciate and exploit the power of computation, biologists must be trained to "think" computationally. In practical terms, this means understanding fundamental computing concepts that recur in many applications *and* being able to write programs.

Why? Consider, for example, that there are well over 400 different software packages for phylogenetics (the study of the evolutionary relationships among organisms). An increasing level of computational sophistication is needed to select the appropriate software for a given application, use it correctly, and understand its abilities and limitations. This is true not just of phylogenetics, but also for many other areas of biology.

But using existing software will not be enough. The number and variety of computational problems that arise in biology are rapidly outpacing the functionality of software tools. Eventually, most biologists are likely to encounter problems that cannot be solved with existing software. Therefore, it is imperative for biologists to have the ability to write their own programs.

This book seeks to provide biology students with both an exposure to major computational ideas *and* practical programming skills. It requires no specific biology or computer science background. It is designed as a first-year college-level course and has been taught to that audience at Harvey Mudd College since 2009. The authors hereby acknowledge the generous grant support of HHMI for the development of that course.

In contrast to a typical introductory bioinformatics book, this book emphasizes programming over the use of existing software. By the time you've completed this book, and the online homework problems, you should feel comfortable writing programs for a wide array of applications in biology and beyond. You'll also have an understanding of computational ideas like "heuristics," "memoization" and "dynamic programming," "NP-completeness," and others, that will allow you to understand and compare the technical aspects of existing bioinformatics tools.

This book begins with Chapter 0, which offers a first introduction to the Python programming language. The rest of the book is organized into four parts, each comprising several chapters. Each part begins with a "large" biological question and the chapters in that part provide the computational and programming tools to answer that question.

At the end of each chapter – and sometimes even within a chapter – you'll see a question icon pointing you to one or more recommended problems. You'll find those problems at the url:

```
www.cs.hmc.edu/CFB
```

Let's get started!

Meet Python

0

One way that you can spot a computer scientist is that they begin counting from 0 rather than from 1. So this is Chapter 0. But it's also Chapter 0 to signify that it's a warm-up chapter to get you on the path to feeling comfortable with Python, the programming language that we'll be using in this book. Every subsequent chapter will begin with an application in biology followed by the computer science ideas that we'll need to solve that problem.

Python is a programming language that, according to its designers, aims to combine "remarkable power with very clear syntax." Indeed, Python programs tend to be relatively short and easy to read. Perhaps for this reason, Python is growing rapidly in popularity among computer scientists, biologists, and others.

The best way to learn to program is to experiment! Therefore, we **strongly urge you to pause frequently as you read this book and try some of the things that we're doing here (and experiment with variations) in Python.** It will make the reading more fun and meaningful.

This chapter includes a number of short exercises that we encourage you to try. Subsequent chapters have links to end-of-chapter programming problems.

The link below offers instructions on how to install and run Python on your computer.

```
www.cs.hmc.edu/CFB/Setup
```

If you're reading this book as a part of a course, your instructor may have some additional or different instructions.

0.1 Getting Started

When you start up Python, you'll see a "prompt" that looks like this:

```
>>>
```

You can type commands at the prompt and then press the "Return" (or "Enter") key and Python will interpret what we've typed.

For example, below we've typed 3 + 5 (throughout this book, the user's input is always shown in black and Python's response is always shown in blue).

```
>>> 3 + 5
8
```

Next, we can do fancier things like this:

```
>>> (3 + 5) * 2 - 1
15
```

Notice that parentheses were used here to control the order of operations. Normally, multiplication and division have higher *precedence* than addition and subtraction, meaning that Python does multiplications and divisions first and addition and subtractions afterwards. So without parentheses we would have gotten...

```
>>> 3 + 5 * 2 - 1
12
```

You can always use parentheses to specify the order of operations that you desire.

Here are a few more examples of arithmetic in Python:

```
>>> 6 / 2
3
>>> 2 ** 5
32
>>> 10 ** 3
1000
>>> 52 % 10
2
```

You may have inferred what /, **, and % do. Arithmetic symbols like +, -, /, *, **, and % are called *operators*. The % operator is pronounced "mod" and it gives the remainder that results when the first number is divided by the second one.

0.2 Big Numbers

Python is not intimidated by big numbers. For example:

```
>>> 2 ** 100
1267650600228229401496703205376L
```

The "L" at the end of that number stands for "Long." It's just Python's way of telling you that the number is big. Look at this:

```
>>> (2 ** 100) % 3
1L
```

The number 1 is *not* long, but 2 ** 100 was long; once Python sees a mathematical expression with a long number in it, it stays in the long number state of mind. You don't have to worry about that – we just note this so you won't say "Huh!?" when you see the "L."

0.3 Strange Division

If you're using Python version 3, division works as you'd expect.

```
>>> 5 / 2
2.5
```

If you're using Python version 2 (e.g., version 2.7), the way it works may surprise you. Take a look at this:

```
>>> 5 / 2
2
```

We've consulted with expert mathematicians and they've verified that 5 divided by 2 is not 2! What's wrong with Python!? The answer is that Python (version 2) is doing "integer division". It assumes that since 5 and 2 are integers, you are only interested in integers. So when it divides 5 by 2, it rounds down to the nearest integer, which is 2. If you wanted Python to do "decimal division", you could do this:

```
>>> 5.0 / 2.0
2.5
```

Since we typed 5.0 and 2.0, Python realized that we are interested in numbers with decimal points, not just integers, so it gave us the answer with a decimal point. In fact, if just ONE of the 5 or 2 has a decimal point after it, Python will get the message that we are thinking about decimal numbers. So we could do any of these:

```
>>> 5.0 / 2
2.5
>>> 5 / 2.0
2.5
>>> 5. / 2
2.5
>>> 5 / 2.
2.5
```

0.4 Naming Things

Python lets you give names to values. This is very useful because it allows you to compute a value, give it a name, and then use it by name in the future. Here's an example:

```
>>> myNumber = 42
>>> myNumber * (10 ** 2)
4200
```

In the first line, we defined myNumber to be 42. In the second line, we used that value as part of a computation. The name myNumber is called a *variable* in computer science. It's simply a name that we've made up and we can assign values to it.

Notice that the "=" sign is used to make an assignment. On the left of the equal sign is the name of the variable. On the right of the equal sign is an expression that Python evaluates, assigning the resulting value to the variable. For example, we can do something like this:

```
>>> pi = 3.1415926
>>> area = pi * (10 ** 2)
>>> area
314.15926
```

In this case, we define the value of `pi` in the first line. In the second line, the expression `pi * (10 ** 2)` is evaluated (its value is 314.15926) and that value is assigned to another variable called `area`. Finally, when we type `area`, Python displays the value. Notice, also, that the parentheses aren't actually necessary here. However, we used them just to help remind us which operations will be done first. It's often a good idea to do things like this to make your code more readable to other humans.

One last thing before we move on. In mathematics, the expression `pi = 3.1415926` is equivalent to the expression `3.1415926 = pi`. In Python, and in almost all programming languages, *these are not the same*! In fact, while Python understands `pi = 3.1415926` it will complain loudly if you type `3.1415926 = pi`. (Try it in Python to see what it looks like when Python complains.) Didn't Python go to elementary school? The answer is that programming languages view the = symbol in a special way. They assume that what's on the left-hand side is the name of a variable (`pi` in our example). On the right-hand side of = is an expression that can be evaluated. That expression is evaluated and its value is then associated with the variable.

So, when Python sees...

```
>>> pi = 3.1415926
```

... it evaluates the expression on the right. That's easy, there's nothing to evaluate – it's 3.1415926. Python then assigns that value to the variable `pi` on the left. Now, when Python sees...

```
>>> area = pi * (10 ** 2)
```

... it evaluates the expression on the right. That's not too hard either; we've already defined `pi` to be 3.1415926 and Python does the math and evaluates the expression on the right to be 314.15926. It then associates that value with the variable named `area`.

Sometimes, you'll see (and write) things like this:

```
>>> counter = 0
>>> counter = counter + 1
```

In the first line, we've set a variable named `counter` to 0. The second line makes no sense from a mathematical perspective, but it makes complete sense based on what we now know about Python. Python begins by evaluating the expression on the right side of the = sign. Since `counter` is currently 0, that expression `counter + 1` evaluates to 1. Only once this evaluation is done, Python sets the variable on the left side of the = sign to be this value. So now `counter` is set to 1. This is how we'll often count the number of events of some sort – like the number of occurrences of a particular nucleotide in a DNA sequence or the number of species with a particular property.

0.5 What's in a Name?

Variable names are pretty much up to you. We didn't have to use the names `myNumber`, `pi`, and `area` in our examples above. We could have called them `Joe`, `Sally`, and `Melissa` if we preferred. Variable names must begin with a letter and there are certain symbols that aren't permitted in a variable name. For example, naming a variable `Joe` or `Joe42` is fine, but naming it `Joe+Sally` is not permitted. You can probably imagine why. If Python sees `Joe+Sally` it will think that you are trying to add the values of two variables `Joe` and `Sally`. Similarly, there are a few built-in Python "special words" that can't be used as a variable name. If you try to use them, Python will give you an *error message*. (An error message is the technical term for a complaint.)

Finally, it's a good practice to use descriptive variable names to help the reader understand your program. For example, if a variable is going to store the area of a circle, calling that variable `area` is a much better choice than calling it something like `z` or `y42b` or `harriet`.

Variables provide a convenient way to store values and the names that we give to variables should be simple and descriptive.

0.6 From Numbers to Strings…

That was all good, but numbers are not the only useful kind of data. Python allows us to define strings as well. A *string is* any sequence of symbols within either single or double quotation marks. Here are some examples:

```
>>> name1 = "Ben"
>>> name2 = 'Jerry'
```

```
>>> name1
'Ben'
>>> name2
'Jerry'
```

Notice that in the first example, we used double quotes around the string `"Ben"` and single quotes around the string `'Jerry'`. Python is happy either way – these are both fine for defining strings. However, notice that Python's own responses (in blue above) use single quotes even if we defined them with double quotes (as in the case of `"Ben"`). That's not a very important detail, but we mention it to avoid any confusion.

A string is any sequence of characters (numbers, letters, and other symbols) beginning and ending with either single or double quotation marks. If the string begins with a single quotation mark it should end with a single quotation mark and if it begins with a double quotation mark it should end with a double quotation mark.

In biology, the term "sequence" generally refers to a sequence of nucleotides or amino acids. Biological sequences are naturally represented by Python strings. Throughout this book, we'll use the terms *string* and *sequence* interchangeably, even though technically one is a computer science concept and the other is a biology concept.

Finding the length of a string

While there are many things that we can do with strings, here are a few of the most important ones. First, we can find the length of a string by using Python's `len` function:

```
>>> len(name1)
3
>>> len('hi there')
8
```

In the first example, `name1` (which we defined above) is `"Ben"` and the length of `"Ben"` is 3. In the second example, `'hi there'` contains a space (which is a regular symbol) so the total length is 8.

Indexing

Another thing that we can do with strings is find the symbol at any given position or *index*. The first symbol in a string actually has index 0 in Python. (As we noted at the beginning of this chapter, computer scientists like to start counting from 0.) So,

```
>>> name1[0]
'B'
>>> name1[1]
'e'
>> name1[2]
'n'
>>> name1[3]
IndexError: string index out of range
```

Notice that although `name1`, which is `"Ben"`, has length 3, the symbols are at indices 0, 1, and 2 by the way that Python counts. There is no symbol at index 3, which is why we got an error message.

0.7 Slicing

The next thing we want to show you is called *slicing*. Python lets you find parts of a string using its special slicing notation. Let's look at some examples first:

```
>>> goodFood = 'chocolate'
>>> goodFood[0:3]
'cho'
>>> goodFood[0:4]
'choc'
```

What's going on here? `goodFood` is a variable representing the string `'chocolate'`. The name of our variable is our own invention – Python has no prior notion of `goodFood` or what it might represent. Now, `goodFood[0:3]` is telling Python to look at the thing which `goodFood` represents (the string `'chocolate'`) and give us back the part beginning at index 0 and going up to, but not including, index 3. So, we get the part of the string with symbols at indices 0, 1, and 2, which are the three letters `'c'`, `'h'`, and `'o'` or `'cho'`.

It may seem strange that the last index is not used – so we don't actually get the symbol at index 3 when we ask for the slice `goodFood[0:3]`. It turns out that there are some good reasons why the designers of Python chose to do this. Humor us for now and we promise to revisit this later.

We don't have to start at 0. For example:

```
>>> goodFood[3:7]
'cola'
```

This is giving us the slice of `'chocolate'` from index 3 up to, but not including, index 7.

We can also do things like this:

```
>>> goodFood[1:]
'hocolate'
```

When we leave the number after the colon blank, it just assumes we mean "go until the end." So, this is just saying, "give me the slice that begins at index 1 and goes to the end." Similarly,

```
>>> goodFood[:4]
'choc'
```

Because there was nothing before the colon, it assumes that we meant 0. So this is the same as `goodFood[0:4]`.

Indexing and slicing allow us to select regions of interest in a string.

0.8 Adding Strings

Strings can also be added together. Adding two strings results in a new string that is simply the first one followed by the second one. This is called *concatenation* of strings. For example:

```
>>> 'yum' + 'my'
'yummy'
```

Once we can add, we can also multiply!

```
>>> 'yum' * 3
'yumyumyum'
```

In other words, `'yum' * 3` really means `'yum' + 'yum' + 'yum'`, which is `'yumyumyum'`; the concatenation of `'yum'` three times.

0.9 Negative Indices

Finally, notice that while it's easy to get the first symbol in a string, it requires a bit more work to get the last symbol. For example, after defining the string

```
goodFood = 'chocolate'
```

we could get the last symbol this way:

```
>>> goodFood[len(goodFood)-1]
'e'
```

This works because the length of the string minus one is the last index of the string, but it's a cumbersome way to index that position! So, Python allows us to get the last symbol in the string using index –1, as in `goodFood[-1]`. Similarly, the second to the last symbol is at `goodFood[-2]`, and so forth.

You can use negative numbers in slicing as well. For example, to get the part of `mystring` that excludes the first and last symbols we could do this:

```
>>> goodFood[1:-1]
'hocolat'
```

0.10 Fancy Slicing

Python's slicing notation also allows you to do things like this:

```
>>> mystring = 'abcdefghijklmnop'
>>> mystring[1:9:3]
'beh'
```

What happened here? The 1 and 9 here mean that we want the slice from index 1 to index 9. In this case that's the string `'bcdefghi'`. The last number, 3, means that we just want every third symbol in that string. So, that is `'b'` (now skip three symbols forward) then `'e'` (now skip three symbols forward) and then `'h'`. There is no symbol 3 symbols forward from `'h'` in the slice `'bcdefghi'`.

Said another way, we got the slice from position 1 to 9 in steps of 3. The steps can also be negative. For example:

```
>>> mystring[5:0:-1]
'fedcb'
```

What's going on now? Here, we asked for the part of the string beginning at index 5 (that's the letter 'f'), to index 0 (that's the 'b' at index 1 because remember that Python stops one step before that index), and going in steps of −1, which means going backwards. Thus, the indices are 5, 4, 3, 2, 1, which is how we got 'fedcb'.

0.11 And Now to Lists...

Sometimes it's convenient to "package" a bunch of things together. Python has a special "container" for data called a *list*.

In Python, a list is simply a collection of things inside brackets like this:

```
>>> odds = [1, 3, 5, 7, 9, 11]
>>> friends = ['Chandler', 'Betty', 'Ross', 'Monica']
```

In the first case, odds is a variable and we've assigned that variable to be a list of six odd numbers. In the second case, friends is a variable and we've assigned that variable to a list of four strings. If you type odds or friends, Python will show you what those values are currently storing.

A list starts with an open bracket [, ends with closed bracket], and each item inside the list is separated from the next item by a comma.

Now, here's something strange:

```
>>> stuff = [2, 'hello', 2.718]
```

Notice that stuff is a variable that is assigned to be a list and that list contains an integer, a string, and decimal number. Python has no objection to different kinds of things being inside the same list! In fact, stuff could even have other lists in it as in this example:

```
>>> stuff = [2, 'hello', 2.718, [1, 2, 3] ]
```

Good news!

Here's some good news: Almost everything that works on strings also works on lists. For example, we can ask for the length of a list:

```
>>> len(stuff)
4
```

Notice that the list `stuff` really only has four things in it: The number 2, the string `'hello'`, the number `2.718`, and the list `[1, 2, 3]`.

Indexing and slicing also work on lists just as in strings:

```
>>> stuff[0]
2
>>> stuff[1]
'hello'
>>> stuff[2:4]
[2.718, [1, 2, 3] ]
```

Imagine that we wanted to get the length of the second item in `stuff` (the word `'hello'`). We could do it this way:

```
>>> len(stuff[1])
5
```

Similarly, we could get the length of the fourth item in `stuff` using:

```
>>> len(stuff[3])
3
```

However, if we ask for `len(stuff[0])` we'll get an error message because that would be asking for the length of the number 2. You might say that the length of the number two is 1, but Python only knows what it means to find the length of a collection of items like a string or a list.

Just like strings, lists can be added and multiplied as well. Adding two lists together creates a new list that contains all of the elements in the first list followed by all of the elements in the second list. This is called *concatenation* of lists and is similar to concatenation of strings.

```
>>> myList = [1, 2, 3]
>>> myList + [4, 5, 6]
[1, 2, 3, 4, 5, 6]
>>> myList * 3
[1, 2, 3, 1, 2, 3, 1, 2, 3]
```

It's worth noting that doing something like. . .

```
>>> myList + [4, 5, 6]
```

. . . doesn't actually change `myList`. Instead, it just gives you a **new** list which is the result of concatenating `myList` and `[4, 5, 6]`. You can verify this by typing myList at the Python prompt and seeing that it wasn't changed.

```
>>> myList + [4, 5, 6]
[1, 2, 3, 4, 5, 6]
>>> myList
[1, 2, 3]
```

0.12 Lists for Free!

Lists are nice, but imagine that you wanted to make the list of all numbers between 1 and 42. That would be a lot of work! Fortunately, Python has a mechanism for making lists of numbers for you. It's called `range` and we'll demonstrate it here. First, here's how it works in Python 2:

```
>>> range(1, 42)
[1, 2, 3, 4, 5, 6, 7, 8, 9, 10, 11, 12, 13, 14, 15, 16, 17, 18, 19, 20,
21, 22, 23, 24, 25, 26, 27, 28, 29, 30, 31, 32, 33, 34, 35, 36, 37, 38,
39, 40, 41]
>>> range(17, 43)
[17, 18, 19, 20, 21, 22, 23, 24, 25, 26, 27, 28, 29, 30, 31, 32, 33,
34, 35, 36, 37, 38, 39, 40, 41, 42]
```

You've probably figured out what's happening here and also noticed something weird: In Python 2, `range(1, 42)` gives us the list of integers from 1 to 41 – one shy of 42. Similarly, `range(17, 43)` gives us the list of integers from 17 to 42 – one shy of 43.

In Python 3, we need just to type `list(range(1, 42))` rather than `range(1, 42)` in order to let Python 3 know that we really want a list.

In general, in Python 2, `range(x, y)` gives us the list of integers from x to y, but stopping one short of y. In Python 3, we use the syntax `list(range(x, y))`.

You might be wondering why `range` stops early. Believe it or not, it's actually a useful thing, as we'll see later.

For now, it's also worth noting that you can do this in Python 2:

```
>>> range(10)
[0, 1, 2, 3, 4, 5, 6, 7, 8, 9]
```

In Python 3, analogously, we do this:

```
>>> list(range(10))
[0, 1, 2, 3, 4, 5, 6, 7, 8, 9]
```

If you give `range` just one number, it assumes that it should start at 0 and go up to one before your number. In other words, `range(z)` is the same thing as `range(0, z)`. And, of course, you can assign a variable to the list that range produces so that you can use that list in various ways later:

```
>>> myList = range(10) # In Python 3, we'd use myList = list(range(10))
>>> myList[3:]
[3, 4, 5, 6, 7, 8, 9]
```

0.13 Changing Values

Let's imagine that we have a variable called `fish` that we'll use to store some number of interest (e.g., the number of fish that we know):

```
>>> fish = 42
>>> fish
42
```

There's nothing surprising so far (other than the fact that we know 42 fish). Now, we meet two more fish and we wish to change the value of our `fish` variable.

```
>>> fish = 44
>>> fish
44
```

Analogously, we can do the same thing with strings or lists or any other kind of data. For example, imagine that we have a variable called `pet` that stores the name of our pet.

```
>>> pet = 'fido'
>>> pet
'fido'
```

Later, we wish to rename our pet. No problem (from Python's perspective)!

```
>>> pet = 'spot'
>>> pet
'spot'
```

And, we could do the same thing for lists. Here's a list named `friends` that stores a list of our friends.

```
>>> friends = ['Sam']
>>> friends
['Sam']
```

Later, we might meet a new friend, Pat, and do this:

```
>>> friends = ['Sam', 'Pat']
>>> friends
['Sam', 'Pat']
```

That probably wasn't surprising. It wasn't our intention to surprise you quite yet! Coming back to our `fish` variable, let's do this:

```
>>> fish = 42
>>> fish = fish + 5
>>> fish
47
```

Let's pause to consider precisely what happened here. First, the variable `fish` is assigned the value 42. In the next line, Python first evaluates the right-hand side of the expression fish + 5. Since fish is 42 at this moment, fish + 5 is 47. Now, the variable fish is changed to take the new value 47.

We can do analogous things with strings:

```
>>> pet = 'fred'
>>> pet = pet + 'dy'
>>> pet
'freddy'
```

And, just to be sure that you're totally happy with this, let's do this one more time for lists:

```
>>> friends = ['Sam', 'Pat']
>>> friends = friends + ['Zephyr']
>>> friends
['Sam', 'Pat', 'Zephyr']
```

In all of these examples we changed the value associated with a variable by *first* computing a new value *and then* assigning our variable to take that new value. In the example above, friends is initially set to be the list ['Sam', 'Pat']. In the next line, Python builds an **entirely new list** friends + ['Zephyr'], which is ['Sam', 'Pat', 'Zephyr']. Finally, Python tells the variable friends "Forget about the old list! Now, your value is this brand new list."

This process of changing a list by *first* computing a brand new list *and then* assigning the variable to this new list as its new value is quite inefficient because making the new list requires copying all of the contents from the original list. Therefore, there's another way to do this in Python and it looks like this:

```
>>> friends = ['Sam', 'Pat']
>>> friends.append('Zephyr')
>>> friends
['Sam', 'Pat', 'Zephyr']
```

Although this appears to be no different from what we did above, there is a subtle difference. When using append, a new list is *not* created. Instead, the starting list is modified (or "mutated") by having a new item added on its end. If you're trying to add something to a large list, this can be advantageous because it's faster.

Lists are said to be *mutable* because they can be changed using the `append` method (and others). Strings and numbers, however, have no such methods. They are *immutable*. As we'll see later, there are also some situations where it's advantageous to be immutable!

0.14 More on Mutability

In Python, mutability can lead to some surprising behaviors. Here's an example:

```
>>> simpsons = ['homer', 'marge', 'bart', 'lisa']
>>> x = simpsons
>>> x
['homer', 'marge', 'bart', 'lisa']
>>> x.append('maggie')
>>> x
['homer', 'marge', 'bart', 'lisa', 'maggie']
>>> simpsons
['homer', 'marge', 'bart', 'lisa', 'maggie']
```

Let's dissect this. We first made a list called `simpsons`. Then, we told Python...

```
x = simpsons
```

... which made `x` a list. Then we changed `x` by appending `'maggie'` to it. When we looked at the `simpsons` list, it had changed too!

The reason for this is that Python, like many other programming languages, is "lazy" when it comes to mutable data, such as lists.[1] When we said...

```
>>> x = simpsons
```

... `x` did not become an entirely new identical copy of the `simpsons` list. Instead, `x` became another name for the `simpsons` list. In other words, the variables `simpsons` and `x` both refer, or "point," to the same list.

Now, when we changed the contents of `x` by mutating that list...

```
>>> x.append('maggie')
```

---

[1] There's some subtle technical stuff that we're finessing, but the discussion here suffices for everything you'll need.

. . . that changed the contents of the list x. But that's the same list as the `simpsons` list, and thus when we asked for the `simpsons` list we got:

```
>>> simpsons
['homer', 'marge', 'bart', 'lisa', 'maggie']
```

The fact that Python does this kind of "lazy" copying is only relevant to us when dealing with mutable things like lists. (In Part II, we'll see one other mutable type of data called a dictionary, but, for now, lists are the only mutable data we've seen.)

In contrast, here's a seemingly similar example:

```
>>> simpsons = ['homer', 'marge', 'bart', 'lisa']
>>> x = ['homer', 'marge', 'bart', 'lisa']
>>> x.append('maggie')
>>> x
['homer', 'marge', 'bart', 'lisa', 'maggie']
>>> simpsons
['homer', 'marge', 'bart', 'lisa']
```

The difference here is that the `simpsons` list and the x list are two separate lists that just happen to look alike. Thus, changing x does not change the value of the variable `simpsons`.

0.15 Booleans

In addition to numbers, strings, and lists, Python has another useful data type called a *Boolean*. A Boolean is just the value `True` or `False`. These are special values and they are not the strings `'True'` and `'False'`.

For example:

```
>>> 5 > 1
True
>>> 3 < 2
False
```

The comparison operators $<$ and $>$ try to compare the quantities on the left- and right-hand sides and, based on that comparison, they tell us `True` or `False`.

To compare two quantities for equality we use == as in:

```
>>> 42 == 2 * 21
True
>>> 'hello' == 'hello'
True
>>> 'hello' == 'goodbye'
False
>>> [1, 2, 3] == [3, 2, 1]
False
```

Notice in that last case, the list [1, 2, 3] is not the same as [3, 2, 1] because they are not in the same order.

A common pitfall is to confuse the = operator with the == operator. Recall that = is used to assign a value to a variable, the variable being on the left of the = and the value being on the right. In contrast, == doesn't assign anything to anyone! It compares the expressions on its left and right and returns the Boolean True or False.

There are also operators >= (greater than or equal), <= (less than or equal), and != (not equal).

A Boolean is a value that is either True or False. A Boolean expression is something that, when evaluated, is either True or False.

You can think of True as meaning "Yes" and False as meaning "No."

Building Boolean expressions with and, or, and not

Finally, we can make interesting Boolean expressions using the Boolean connectors, or *operators*, and, or, and not as in:

```
>>> 2 > 1 and 3 < 4
True
>>> 'hello' == 'goodbye' or 3 == 1+2
True
```

You can think of and and or as kind of Boolean "glue" that glues two Boolean expressions together to form a new one. Both of these operators expect the things

© S. Harris/http://www.cartoonstock.com.

on their left and right sides to be Boolean expressions. For example, in the and case above, 2 > 1 is a Boolean expression: its value is True. Also, 3 < 4 is a Boolean expression: its value is also True. Now, the and says, "if the expression on my left is True **and** the expression on my right is True, then I'll be True. But in any other case, I'll be False." Since the expression 2 > 1 is True and the expression 3 < 4 is True, the expression 2 > 1 and 3 < 4 is True.

The or "glue" is similar, but it's True if *at least one* of the expressions on its left or the expression on its right (or both) are True. Although 'hello' == 'goodbye' is False, 3 == 1+2 is True. That's all we need for the expression 'hello' == 'goodbye' or 3 == 1+2 to evaluate to True.

There's also something called `not`. This simply reverses (or "negates") the value of the expression to its right. For example

```
>>> not 3 == 4
True
```

Notice that `3 == 4` is `False`. Therefore, `not 3 == 4` is `True`.

Putting it All Together

Whew! In this chapter we got a whirlwind taste of Python's basic types of data: Numbers, strings, lists, and Booleans. We explored some of the ways that these data can be manipulated and we saw that we can give names to data using variables. In the next chapter we'll take an important next step: learning how to write programs that manipulate data in interesting and useful ways.

PART I
Python versus Pathogens

· ·

In 1906, the Warrens, a wealthy New York banking family, rented a summer house on Long Island. That summer, six people in the household came down with typhoid fever, a serious bacterial illness with a persistent and very high fever. In the era before antibiotics, typhoid was frequently deadly.

Although the Warrens all survived, the outbreak was troubling enough that a sanitary engineer named George Soper was hired to investigate. Soper examined the water supply, the plumbing, and other possible sources of contamination, but found nothing to explain the outbreak. Eventually, he investigated the family's new cook, a woman named Mary Mallon. Soper went through her employment history, and found that there had been typhoid outbreaks in most of the places she had worked. Mary Mallon, who became known as "Typhoid Mary," was the first documented example of an asymptomatic carrier of typhoid (Figure 1.1). She herself was not sick, but she was able to spread the disease to others. Once Mary was discovered to carry typhoid, she was quarantined in a hospital for most of the rest of her life.

Typhoid fever is caused by a bacterium called *Salmonella enterica typhi*. An infection begins when contaminated food or drink is ingested, and the bacteria are able to make their way to the gut. Once there, they invade the cells lining the intestine. From these cells, they are able to spread to other parts of the body. This leads to a serious systemic infection that, if left untreated, is fatal in roughly 10–30% of cases. Among those who survive an untreated infection, a small number become carriers like Mary Mallon. In these cases the bacterium is able to persist in the body even after the main infection has been cleared by the immune system.

The *Salmonella* strain that causes typhoid fever has some close relatives that also cause disease. Outbreaks of food poisoning are often due to a strain called *Salmonella enterica typhimurium*. The two strains of *Salmonella* share a number of characteristics, including the basic mechanism of infecting gut epithelial cells.

Over the next three chapters, we'll explore this mechanism using computational techniques. The ultimate goal in these chapters is to write a computer program that will analyze a DNA sequence, such as that of a given strain of *Salmonella*, and identify its genes. Once your program has found the genes, you'll use an existing database to compare them to genes that have been studied previously and whose function is known. Ultimately, this will lead you to identify pathogenic genes in *Salmonella* and determine their function. This type of analysis is useful in studying diseases of all sorts.

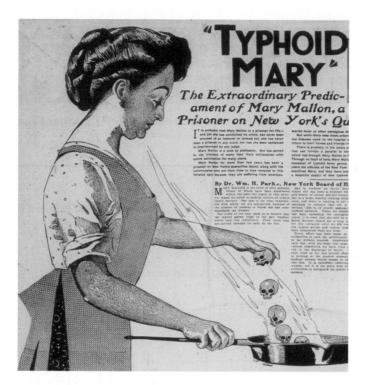

Figure 1.1. A 1909 article about Typhoid Mary from the *New York American* (http://www.sciencemuseum.org.uk/broughttolife/people/typhoidmary.aspx).

Computing GC Content

Computing content:
functions,
conditionals (if/elif/else),
for loops over strings

Nucleic acids are the chief information-bearing molecules in cells. Recall that they consist of polymers of repeating units called *nucleotides* that contain a sugar, a phosphate group, and a nitrogenous structure called a *base*. The base is the part that is variable and thus carries information. In deoxyribonucleic acid (DNA) there are four types of bases: adenine (A), cytosine (C), guanine (G), and thymine (T). The sequence of a nucleic acid polymer is defined by the order of these bases, which we can represent with a string of A's, C's, G's, and T's.

Large strands of DNA, such as are found in bacterial chromosomes, can have millions of nucleotide units. Though you might expect that the four bases would occur in roughly equal numbers in such sequences, this is often not the case. The percentage of nucleotides that are G or C, called the *GC content*, varies considerably among organisms and can be used to categorize and compare them. For example, *Salmonella enterica typhi*, the pathogenic species of *Salmonella* that causes typhoid fever, has a GC content of approximately 52%. The GC content of other bacteria ranges from about 25% to 75%.

In fact, GC content can even vary from one part of the genome to another within a single species. In particular, genomic regions with pathogenic genes frequently have a different GC content than the rest of the genome. As we'll see, this feature can be very useful in identifying genomic regions that have pathogenic genes.

1.1 Representing DNA on a Computer

The DNA molecules found in cells consist of two strands that wrap around one another in a so-called "double helix." The two strands are linked to one another via hydrogen bonds with adenine (A) bases bonding to thymine (T) and cytosine (C) bonding to guanine (G). These are called complementary *base pairs* (bp). We'll use the term "bp" quite a lot in this chapter.

Each strand of DNA has an orientation defined by the sugars in its nucleotides. The two ends of a strand are called the 5′ and 3′ ends. While one strand goes 5′ to 3′, the other strand goes 3′ to 5′. One way to represent such molecules is to include both strands:

```
5' - ATTCGTCA - 3'
3' - TAAGCAGT - 5'
```

Since the strands are complementary, one strand can be deduced from the other. For this reason, computer representations of DNA typically only show one of the two strands. The convention is to write the sequence in the 5′ to 3′ direction. Thus, the above sequence could be written either like this:

```
ATTCGTCA
```

or like this:

```
TGACGAAT
```

These two strings are called *reverse complements*. Note that their nucleotides are complementary (A's interchanged with T's and C's interchanged with G's) and in the reverse order.

Our goal for this chapter is to write a program that computes the GC content of a DNA string. Note that for every G or C on the forward strand, there is a corresponding C or G on the reverse strand. So, the GC content of the forward strand is the same as the GC content of the reverse strand and it suffices to compute the GC content of just one of the two strands. We'll work with the forward strand.

1.2 Python Functions

A program is similar in many ways to the idea of a function in mathematics. In math, a function can be thought of as a "box" that takes some data as input and returns some data as output. For example, the function $f(x) = 2x$ has an input called x and for any value that we "stick into" the input we get back an output value that is twice as large.

Python lets us define functions too and, as in math, these functions take some data as input, process that data in some way, and then return some data as output. Here's a Python function that we've named f that takes in an input which we are calling x and returns an output that is two times x:

```
def f(x):
    return 2 * x
```

In Python, the syntax for functions uses the special word def which means: "I'm defining a function." Then comes the name that we've chosen for the function; in our case we chose to call it f. Then comes the input to the function in parentheses (just like the definition of a function in math!), followed by a colon (":"). Then, we start a new line, indented a few spaces (typically a tab stop), and begin the contents of the function. In this case, our function computes 2 * x and then returns it. The word return tells Python to give us back that value.

Don't forget to indent the line (or lines) after the first line. Python absolutely demands this. We'll talk a bit more about indentation shortly.

We can now use, or *run*, the function in the Python interpreter like this:

```
>>> f(10)
20
>>> f(-1)
-2
```

When we type f(10) we say that this is a *function call* because we are metaphorically placing a call to our function f with input 10 and asking it to give us an answer. When we call f, Python takes the value inside the parentheses and assigns it to the name x. Then it executes the statement or statements inside the function in order. In this case, there is only a single statement inside the function; this statement doubles the value of x and returns that value back to us.

As we noted, you can name a function more-or-less anything you want (as with variable names, there are a few exceptions but it's not worth enumerating them). In this example, we named the function f simply to make the analogy with the mathematical function $f(x) = 2x$ that we defined earlier. In general, it's better to give functions descriptive names, just as we've advocated doing for variables. So, calling the function something like double would probably have been a better choice.

We've just written our first program!

A program is one or more functions that work together to solve a problem.

Our example program has just one function and, admittedly, it's nothing to write home about – but it's a good first step. Soon, we'll write programs that involve multiple interacting functions that do more interesting things.

1.3 A Short Comment on Docstrings

As our programs become more interesting and complex, it will be important for users of the program to be able to quickly understand what the program does. To that end, every function that we write from now on will begin with something called a *docstring*, which stands for "*do*cumentation *string*." Here's our snazzy doubling function f with its docstring:

```
def f(x):
    ''' Takes a number x as input and returns 2 * x.'''
    return 2 * x
```

The docstring is a string that begins and ends with three single quote marks and describes what the function does. We'll see many examples of this in the remainder of this chapter and the book. The nice thing about docstrings is that you if you type help(f) the docstring for function f will be displayed.

You should always provide a docstring for every function that you write.

1.4 Bigger and Better Functions

Functions can have more than one line

The function above, f, contained only one statement: return 2 * x. However, in general, functions are not limited to a single Python statement, but rather may have as many statements as you wish. For example, we could have written the function this way:

```
def f(x):
    ''' Takes a number x as input and returns 2 * x.'''
    twoTimesX = 2 * x
    return twoTimesX
```

Notice that when there is more than one statement in the function they must all be indented to the same position. This is how Python knows which statements are inside the function and which are not. We'll say even more on indentation in a moment!

Here you might be wondering which definition for f is better: the one with one line or the one with two. The answer is that both are equally valid and correct. For this relatively simple function, most programmers would use the one-line version.

Functions can have multiple inputs

Functions can have more than one input. For example, here's a function that takes two inputs, x and y, and returns $x^2 + y^2$:

```
def sumOfSquares(x, y):
    ''' Computes the sum of squares of its two inputs.'''
    return x ** 2 + y ** 2
```

1.5 Why Write Functions?

At this point you might be wondering why we even write functions. Multiplying a number by 2 or computing the sum of squares of two numbers hardly seems to justify a computer program!

But functions often do much more complex computations than just doubling their input. Generally, functions allow us to "package" a sequence of computations that we know we will want to perform frequently. Then, we can assemble those packages into larger programs that do some truly interesting and useful things.

1.6 Making Decisions

Our goal for this section is to compute the GC content of a DNA string. For the moment, let's imagine that we know in advance that the DNA string will be exactly four bp long, as in 'AAAT' or 'CGCT'.

Let's name our function GCcontent4 (the 4 indicating that this only works for DNA strings of length 4). Ultimately, we'd like to use our GCcontent4 function like this:

```
>>> GCcontent4('AGAC')
0.5
>>> GCcontent4('AAAC')
0.25
```

Notice that we're getting the GC content back as a number between 0 and 1 (rather than a percentage). Also notice that this function is taking a *string* as input rather than a *number*. That may seem a little weird since the mathematical functions that we're most familiar with take numbers as input and produce numbers as output. Python is happy to have its functions take any kind of input and produce any kind of output.

If we were to compute the GC content by hand, we'd need to somehow keep count of the number of G's and C's. To that end, imagine that we have a "counting cup" that starts off empty. We begin by looking at the first symbol in the string. If it's a G or a C, we place a pebble in the counting cup to indicate that we found one of our desired symbols. Then, we move on to the next symbol in the string and check to see if it's a G or a C. If it is, we add another pebble to the cup. We do the same thing for the third and fourth symbols. Finally, we look at the number of pebbles in the cup and divide that number by four (the length of the input string) to get our GC content.

Here's a function that does exactly this, using numbers rather than pebbles. It uses a few things that we haven't seen before, but it will probably make sense anyhow. Take a look and we'll dissect this together in a moment. (Every biology book should have some dissection, after all!)

```
def GCcontent4( DNA ):
    ''' Computes the GC content of a DNA string of length 4.'''
    counter = 0
    if DNA[0] == 'G' or DNA[0] == 'C':
        counter = counter + 1
    if DNA[1] == 'G' or DNA[1] == 'C':
        counter = counter + 1
    if DNA[2] == 'G' or DNA[2] == 'C':
        counter = counter + 1
    if DNA[3] == 'G' or DNA[2] == 'C':
        counter = counter + 1
    return counter / 4.0
```

What's going on here? First, the name that we've given to the input to this function is DNA rather than x. We could have called it anything, but DNA is a descriptive name for this variable. Now, when we call this function as in...

```
>>> GCcontent4('CGAC')
```

...the variable DNA will take the value 'CGAC'.

Inside this function, we have a variable named counter that is playing the role of our metaphorical counting cup. The first line of our function sets that variable to have the value 0 since we haven't seen any G's or C's yet.

Take a look at the next two lines of this function:

```
if DNA[0] == 'G' or DNA[0] == 'C':
    counter = counter + 1
```

Here we are using the string indexing notation that we saw in Chapter 0 to look at the first symbol in the DNA string to see if it's either a G or a C. (Recall that DNA[0] refers to the first position in the DNA because Python follows the convention of counting from 0.) If that first symbol is a G or a C, we increment the counter by 1. Then, we move on to the next lines of the program where we do this same thing for the next index in the DNA string and so forth.

A bit more dissection of this function is warranted. How exactly does that if business work? Recall from Chapter 0 that the expression DNA[0] == 'G' is a Boolean expression: It's either True or False. Similarly, the expression DNA[0] == 'C' is a Boolean expression. Therefore, the expression DNA[0] == 'G' or DNA[0] == 'C' is a Boolean expression because the or works like Boolean "glue" on the expressions to its left and right. For example, if DNA is 'CGAC', then DNA[0] is C. So, in this case DNA[0] == 'G' is False and DNA[0] == 'C' is True. Therefore, the expression

```
DNA[0] == 'G' or DNA[0] == 'C'
```

...is True. In Python, the if statement, also called a *conditional statement*, works like this: What comes after the if is a Boolean expression followed by a colon. Python will evaluate that Boolean expression and, if it's True, Python will execute the commands that are indented on the subsequent lines. In our case, there's just one indented line after the if, and that line serves to increment the variable counter by 1 to account for the fact that we've found a G or a C.

Now take one more look at our GCcontent4 function. Notice that after initializing the variable counter to 0 it uses an if statement four times, once for each of the four positions in the DNA string. If a position contains a G or a C, the counter will be incremented by 1. If not, the function will move on to the next if statement. At the end of the four if statements, the counter will have the value 0, 1, 2, 3, or 4 depending on how many positions had a G or a C. Dividing the value in counter by 4.0 gives us the GC content.

1.7 A Potential Pitfall

Our function contains an if statement that looks like this:

```
if DNA[0] == 'G' or DNA[0] == 'C':
    counter = counter + 1
```

What if, instead, we'd used the following?

```
if DNA[0] == 'G' or 'C':
    counter = counter + 1
```

Strangely, this does not work as intended even though it seems simpler and more natural. What's wrong then? Recall that or (like its sibling, and) expects that what's to its immediate left and right are both Booleans. For example, 1 == 2 or 2+4 == 6 makes sense because the expression 1 == 2 evaluates to False (a Boolean) and the expression 2+4 == 6 evaluates to True (also a Boolean). Thus we have Booleans on both sides.

On the other hand, in the line...

```
if DNA[0] == 'G' or 'C':
```

...what's to the left of the or is DNA[0] == 'G', which is certainly a Boolean (it's either True or False depending on the the value of DNA[0]). However, what's on the right of the or is simply 'C'. And, 'C' is a string and not a Boolean. That's the problem!

What if my string isn't of length 4? if and else

Our GCcontent4 function assumes that the input DNA has length exactly 4. That's not a reasonable assumption in general and we'll try to do better. As a very modest first step, consider this variant for strings of length 3 or 4. (That's still not so useful since we'd like to handle strings of any length, but we'll get there soon!)

```
def GCcontent3or4( DNA ):
    ''' Computes the GC content of a DNA string of length 3 or 4.'''
    counter = 0
    if len(DNA) == 3:
        if DNA[0] == 'G' or DNA[0] == 'C':
            counter = counter + 1
        if DNA[1] == 'G' or DNA[1] == 'C':
            counter = counter + 1
        if DNA[2] == 'G' or DNA[2] == 'C':
            counter = counter + 1
        return counter / 3.0
    else:
        if DNA[0] == 'G' or DNA[0] == 'C':
            counter = counter + 1
        if DNA[1] == 'G' or DNA[1] == 'C':
            counter = counter + 1
        if DNA[2] == 'G' or DNA[2] == 'C':
            counter = counter + 1
        if DNA[3] == 'C' or DNA[3] == 'C':
            counter = counter + 1
        return counter / 4.0
```

There are a few new things here; let's walk through this. Imagine that we call this function with an input string of length 3. The counter starts at 0. If the length of the DNA string is 3, then when Python gets to the line...

```
if len(DNA) == 3:
```

... it sees that the Boolean expression `len(DNA) == 3` is `True`. It therefore executes all of the lines indented below this `if` statement. That set of indented lines is called a "block." In this case, the block is:

```
        if DNA[0] == 'G' or DNA[0] == 'C':
            counter = counter + 1
```

```
    if DNA[1] == 'G' or DNA[1] == 'C':
        counter = counter + 1
    if DNA[2] == 'G' or DNA[2] == 'C':
        counter = counter + 1
    return counter / 3.0
```

If the length of the DNA string is indeed 3, then at the end of this block we return the value `counter / 3.0`. The `return` statement is very strong; it gives us back that value and terminates execution of the function. The fact that there is more stuff below is irrelevant to Python – `return` means "done!"

On the other hand, if the length of the DNA is 4, then the Boolean `len(DNA) == 3` is `False`. Therefore, when Python reaches the line...

```
if len(DNA) == 3:
```

... it does *not* enter the block immediately below it. Instead, it skips to the next line at the same level of indentation. In this case, it's the short line

```
else:
```

The meaning of this `else` is simply this: Because the `if` statement didn't enter its own block, we instead execute the block immediately below the `else` statement. In this case, it's the same sequence of steps that we did earlier to compute the GC content of a string of length 4.

1.8 GC Content of Strings of Length 1, 2, 3, or 4: `if, elif, else`

Let's take this one step further and write our GC content function so that it works for strings of length 1, 2, 3, or 4. That's progress, but hardly perfect (what about strings of length 5, 6, etc.?). We promise that perfection is just around the corner!

```
def GCcontent1thru4 ( DNA ):
    ''' Computes GC content of a DNA string of length 1 thru 4.'''
    counter = 0
    if len(DNA) == 1:
        if DNA[0] == 'G' or DNA[0] == 'C':
            counter = counter + 1
        return counter / 1.0
```

```
    elif len(DNA) == 2:
        if DNA[0] == 'G' or DNA[0] == 'C':
            counter = counter + 1
        if DNA[1] == 'G' or DNA[1] == 'C':
            counter = counter + 1
        return counter / 2.0
    elif len(DNA) == 3:
        if DNA[0] == 'G' or DNA[0] == 'C':
            counter = counter + 1
        if DNA[1] == 'G' or DNA[1] == 'C':
            counter = counter + 1
        if DNA[2] == 'G' or DNA[2] == 'C':
            counter = counter + 1
        return counter / 3.0
    else:
        if DNA[0] == 'G' or DNA[0] == 'C':
            counter = counter + 1
        if DNA[1] == 'G' or DNA[1] == 'C':
            counter = counter + 1
        if DNA[2] == 'G' or DNA[2] == 'C':
            counter = counter + 1
        if DNA[3] == 'G' or DNA[3] == 'C':
            counter = counter + 1
        return counter / 4.0
```

The one new twist here is the introduction of the `elif` statements. You should think of `elif` as being pronounced "else if" – it's a sort of combination of `else` and `if`. The first `if` statement checks to see if the DNA length is 1. If so, it executes its block and returns the GC content. If the length of the DNA is not 1, then this block is skipped and Python proceeds to the next statement at the same level of indentation as that first `if` statement. That's the line:

```
elif len(DNA) == 2:
```

This line checks to see if the length of the DNA is 2. If so, it executes its indented block and returns the GC content. Otherwise, it drops down to the next `elif` statement, which checks whether the length of the DNA string is 3. Finally, if the length of the DNA is not 3 either, then it executes the `else` statement. The `else` is a catch-all for all remaining cases not "caught" by the first `if` or the subsequent `elif`s.

In general, the structure of an `if` statement works like this: The `if` statement tests whether some Boolean expression is `True`. If so, we execute the block of code indented beneath the `if` statement. When that block is done, we skip any `elif` and `else` statements and proceed to the next line at the same level of indentation as our `if` statement.

   Optionally, we may have some number of `elif` statements following the `if`. The first `elif` statement whose Boolean expression is `True` gets to execute its block of code. Again, once that block is done, we're done with this `if/elif/else` sequence and skip any remaining `elif` and `else` statements, proceeding to the next line at the same level of indentation as our `if` statement.

   Finally, we can optionally have an `else` statement at the end. In other words, the `else` is not required but if we have one we can have only one. The `else` serves as a catch-all and it has no associated Boolean expression that it tests. We simply execute its block if none of the Booleans in the `if` or following `elif`s evaluate to `True`.

1.9 Oops!

One problem with our `GCcontent1thru4` function is that it returns an answer, but potentially the wrong answer, when the DNA string has length more than 4. How so? Imagine that we use a DNA string whose length is 1000 in our function. And imagine that the string begins with AAAA and then has 996 G's following it. Our function will report a GC content of 0 when in fact that GC content is 0.996. Why?

   Since the length of the string is not 1, 2, or 3, the `else` block is executed. That block works perfectly fine, but it only counts the number of G's and C's in the first four positions of the DNA string. If the length of the string is greater than 4, we should warn the user that the length is not in the range that we're able to deal with and return a suitable message.

   In general, we must be careful to return an appropriate message rather than a value that the user might interpret incorrectly.

Here's one possible way to do this:

```python
def GCcontent1thru4 ( DNA ) :
    ''' Computes GC content of a DNA string of length 1 thru 4.'''
    counter = 0
    if len(DNA) == 1:
        if DNA[0] == 'G' or DNA[0] == 'C':
            counter = counter + 1
        return counter / 1.0
    elif len(DNA) == 2:
        if DNA[0] == 'G' or DNA[0] == 'C':
            counter = counter + 1
        if DNA[1] == 'G' or DNA[1] == 'C':
            counter = counter + 1
        return counter / 2.0
    elif len(DNA) == 3:
        if DNA[0] == 'G' or DNA[0] == 'C':
            counter = counter + 1
        if DNA[1] == 'G' or DNA[1] == 'C':
            counter = counter + 1
        if DNA[2] == 'G' or DNA[2] == 'C':
            counter = counter + 1
        return counter / 3.0
    elif len(DNA) == 4:
        if DNA[0] == 'G' or DNA[0] == 'C':
            counter = counter + 1
        if DNA[1] == 'G' or DNA[1] == 'C':
            counter = counter + 1
        if DNA[2] == 'G' or DNA[2] == 'C':
            counter = counter + 1
        if DNA[3] == 'G' or DNA[3] == 'C':
            counter = counter + 1
        return counter / 4.0
    else:
        return 'I cannot handle strings of this length.'
```

Here we have an `if` and three `elif`s followed by a single `else`. The `else` gets invoked when the length of the string is not 1, 2, 3, or 4.

1.10 The "Perfect" GC Content Function: `for` Loops

Our most recent function for computing the GC content worked just fine assuming we knew that the DNA string had length 1 through 4. But trying to extend that idea to work for strings of any length was clearly going to be a problem. After all, there are an infinite number of different string lengths and we can't have a special case for each of them! (That would, quite literally, take forever.)

Fortunately, there's a way of dealing with this problem using a new ingredient called a `for` loop. A `for` loop allows Python to march through every character in our string and do the "same thing" with each character, no matter how long that string might be.

Here's an example of a function with a `for` loop. We'll show it to you first and then explain what's happening.

```
def length( myString ):
    ''' Computes the length of its input string.'''
    counter = 0
    for myCharacter in myString:
        counter = counter + 1
    return counter
```

Let's imagine that we run this function with the input string `'CAT'` so the variable `mystring` refers to the string `'CAT'`. The function begins by setting the variable named `counter` to 0. Next, it reaches the line:

```
    for myCharacter in myString:
```

We know that `myString` is the string `'CAT'` in this case. What about that name `myCharacter`? That's simply a variable – we could have called it anything but this name is a pretty good one. Here's what Python will do with it: First, it will assign `myCharacter` to be the first character in the string referred to by `myString`, that is, `'C'`. Now, Python goes to the indented line...

```
    counter = counter + 1
```

... thereby incrementing `counter` from 0 to 1. Notice that this is the only line in the block for this `for` statement because it's the only line that's indented below it. The line below that one, containing the `return` statement, is *not* indented with respect to the `for`. So, Python executes the line `counter = counter + 1` and the block is done.

Now, Python goes back to the `for` statement, but this time it sets `myCharacter` to be the next letter in the string `myString`. In this case it's the letter `'A'`. Python enters the block for this loop again, thereby incrementing `counter` again. Now `counter` has value 2. The block is complete and we come back to the `for` loop. This time, the variable `myCharacter` becomes the letter `'T'`. So, `counter` is incremented again, this time to 3. Finally, when Python comes back to the `for` loop, it sees that `myCharacter` has taken each value in `myString`, from the first to the last, and the loop is done.

At this point, Python continues to the first line *after* the block – that is, the next line that is indented at the same level as the `for` loop. In this case, that's the `return` statement where the value of `counter` is returned. Notice that this function simply computes and returns the length of the input string. It's doing the same thing as the built-in `len` function that we saw in Chapter 0. We did this just to show you how `for` loops work. Next, we'll do some more sophisticated things.

1.11 Another Example of `for` Loops: Converting DNA to RNA

Let's take a look at another example. You may recall that RNA has the same nucleotides as DNA except it uses uracil (U) instead of thymine (T). So, for example, for the DNA string `'TTACG'`, the equivalent RNA would be `'UUACG'`. (Indeed, as you may know, the process of transcription in a cell involves creating an RNA sequence that is equivalent to one strand of the double-stranded DNA molecule.) Here's a function that takes a DNA string as input and returns the corresponding RNA:

```
def DNAtoRNA ( DNA ) :
    ''' Returns the equivalent RNA for the given DNA string.'''
    RNA = "
    for nuc in DNA:
        if nuc == 'T':
            RNA = RNA + 'U'
        else:
            RNA = RNA + nuc
    return RNA
```

This function takes a string, which we've called DNA, as input. It makes a new string, which we've called RNA, which is initially the empty string. (This is analogous in spirit to setting our counter to 0 in the length function. In both cases, we plan to build a result that will ultimately be returned.) In order to build our RNA string, we need to march through the DNA string and add each character in the DNA string to the RNA string. However, if the DNA character happens to be a 'T', we change it to a 'U'.

There are a few important things to notice in this function. First, we're updating a *string* rather than a *number* as we did in the length function. For example, the line RNA = RNA + nuc takes the current value of RNA (it's initially the empty string), adds the character nuc to the end of it and, after constructing this new string, assigns RNA to refer to that new string.

The other thing to notice here is that the block below the for loop is more than one line long – it's now all four lines indented below it. It might seem strange that the block contains conditional statements, but that's fine: The block can have anything we wish inside it (even other for loops – but more on that later).

In general, a for loop looks like this:

```
for variableName in stringName:
    block
```

The variableName is any name that you choose, stringName refers to some existing string, and block is a collection of one or more lines that get executed as the variable named variableName takes turns marching through the characters in stringName.

Putting it All Together

In this chapter we've explored the basics of programming in Python. We've seen functions (which are essentially short programs), conditional statements (if, elif, else), and for loops. Your final task for this chapter is to write a function to compute the GC content of a DNA string. In the next chapter, you'll use this function to write a program that identifies putative pathogenicity islands (regions with many disease-causing genes) by identifying areas of the genome with anomalous GC content.

Pathogenicity Islands

Computing content:

for loops over lists,
iterating over indices of a string,
composition of functions

Certain regions of the *Salmonella* genome contain genes that are directly involved in causing disease. These so-called *pathogenicity islands* have genes with functions related to invading and living inside a host organism. As you might expect, medical researchers are very interested in identifying and studying such regions. One characteristic that is useful in locating pathogenicity islands is GC content. The GC content inside a pathogenicity island often differs significantly from what is found in the rest of the genome. For this reason, it's useful to find sections of the genome with unusual GC content. This allows us to zoom in on parts of the genome that are candidate pathogenicity islands.

In this chapter, you'll write a program that computes and reports the GC content of different regions of the genome. In the following chapters, we'll refine our search even further to find individual genes in the pathogenicity islands.

2.1 How *Salmonella* Enters Host Cells

Below is an image showing *Salmonella* bacteria invading cultured human cells. The bacteria will physically enter and reside inside these human cells.

Salmonella have evolved an elaborate mechanism that tricks host cells into letting them in. In Figure 2.1, the waviness of the human cell membranes, called

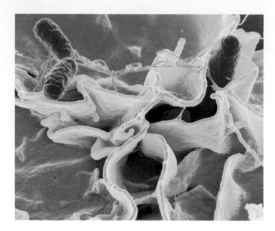

Figure 2.1. A colorized scanning electron micrograph showing *Salmonella* bacteria in red and cultured human cells in green. The bacteria are invading the human cells. Image credit: NIAID, NIH, USA (http://www.niaid.nih.gov/topics/biodefenserelated/biodefense/publicmedia/Pages/image_library.aspx).

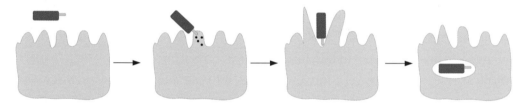

Figure 2.2. Schematic of *Salmonella* entering a host cell.

membrane ruffling, is a characteristic that has been induced by the bacteria to aid their entry.

The bacteria make use of a specialized protein "machine" that is able to inject other proteins into the host cells. The process is illustrated in Figure 2.2. The bacteria (magenta) first attach to the surface of host cells (gray) lining the intestine, and then use their injector system (cyan) to insert certain proteins into the host cell. These proteins hijack various host cell processes, ultimately causing the bacteria to be engulfed in a membrane-bound vesicle, and brought into the cell.

Human and other eukaryotic cells have an internal structural support system called the *cytoskeleton.* During an infection, *Salmonella* "hijacks" the regulation of some proteins in the cytoskeleton. This is how the bacteria are able to get the host cell to produce membrane ruffling, and ultimately how they get themselves engulfed and brought into the host cell.

2.2 Investigating Pathogenicity Islands

Computational methods played a big role in helping scientists determine how the *Salmonella* host cell entry system works. In particular, computer programs were helpful in studying the genomic region that codes for the injector. As is often the case, the first step was in the laboratory. It involved a comparison between the *Salmonella* genome and that of *Escherichia coli* (a related bacterial species that doesn't have the injector mechanism).

This research began in the 1980s, and at that time it was not yet possible to sequence whole bacterial genomes. However, there were laboratory methods available to identify regions of DNA that are unique to one species or another. Using these techniques, a number of regions were found that are present in *Salmonella* but not *E. coli*.

This is where the computational part of the story begins. A laboratory studying one of these regions decided to determine its sequence. This particular region, called RF319, was about 6500 bp long and was known to contain a number of genes. The researchers needed to find where each of those genes started and ended, and they needed to determine their function. It turns out both of these problems can be addressed computationally. In subsequent chapters we'll return to these questions. For now let's skip to the results.

Research into the RF319 region found that it coded for a number of genes related to host cell entry in *Salmonella*. Subsequent studies have shown that RF319 is part of a larger region, approximately 40,000 bp long, that contains genes for the injector mechanism. This larger region is what is known as a *pathogenicity island* – a part of a bacterial genome containing genes involved in infection and disease. Figure 2.3 depicts this particular pathogenicity island.

In subsequent chapters we'll return to the problem of identifying genes and determining what they do. In this chapter, we'll write programs to identify likely pathogenicity islands by looking for regions whose GC content differs from that of the overall genome.

Figure 2.3. The 40 kb region of *Salmonella* pathogenicity island 1. Genes are represented by magenta or blue boxes. Magenta represents genes in the pathogenicity island. These are found in *Salmonella* but not in *E. coli*. Blue represents genes which lie outside the pathogenicity island, and which can be found in both *Salmonella* and *E. coli*.

2.3 Looping Over Lists

In Chapter 1 we saw how `for` loops can be used to iterate over the characters in a string. In fact, `for` loops can be used in much the same way to iterate over the elements in a list. After all, strings and lists are very similar; for example, recall from Chapter 0 that both lists and strings use the same notation for indexing and slicing.

Here's a function that takes a list of numbers as input and returns the sum of the numbers in that list:

```
def add( myList ):
    ''' Returns the sum of the elements in the given list.'''
    sum = 0
    for num in myList:
        sum = sum + num
    return sum
```

We can run this function like this:

```
>>> add([0, 1, 2, 3])
6
```

There's really nothing new here, except that in the statement...

```
for num in myList:
```

... the variable that we've named `num` is, one-by-one, taking the values in `myList`, which happens to be a list. In our example we passed the list `[0, 1, 2, 3]` into the input variable `myList`. In this case `num` was first 0, then 1, then 2, and finally, 3.

The general pattern for a `for` loop over a list is just like the one we saw for strings:

```
for variableName in list:
    block
```

The variable named `variableName` takes each value from `list`, one at a time. Each time, it enters the block and performs the instructions there. This continues until `variableName` has taken the value of each element in `list`.

2.4 Looping Over Lists with `range`

According to legend, when the great mathematician Carl Friedrich Gauss misbehaved in elementary school, his teacher gave him the unpleasant task of adding up the numbers from 1 to 100. This was in the late 1700s and Gauss didn't have a computer. Instead he came up with an ingenious way to find the sum.

We'll do it in Python. We could use our `add` function that we wrote a moment ago, but it would be a nuisance to type in the list of numbers from 1 to 100. You might recall the `range` function from Chapter 0. It builds lists of integers for us. So, we could add up the integers from 0 to 100 like this:

```
>>> myNumbers = range(101)
>>>  add(myNumbers)
5050
```

Alternatively, we could write a new function called `addUpTo` that takes a single number, call it `largest`, that adds up the numbers from 0 to `largest`. We would write it this way:

```
def addUpTo( largest ):
    ''' Returns the sum of the integers from 0 to largest.'''
    sum = 0
    for num in range(0, largest+1):
        sum = sum + num
    return sum
```

Now, we could run it like this:

```
>> addUpTo(100)
5050
```

Based on what we've said so far about Python 3, you'd expect that we would need to replace the line...

```
for num in range(0, largest+1):
```

with the line...

```
for num in list(range(0, largest+1)):
```

However, in Python 3 both versions above will work: In a `for` loop, Python 3 doesn't require the word `list` in front of `range`. So, from here on out, our `for` loops will omit the word `list` and the code will work in both Python 2 and 3.

Don't chew gum in class

We're not sure what got young Gauss in trouble with his teacher. But we do know (from personal experience) that chewing gum is one way to get in trouble. Fortunately, a computer program can help if you're caught! Here's a program that takes a number as input – we've named it `times` – and prints the string `'I will not chew gum in class'` that many times.

```
def gum( times ):
    ''' Apologizes for gum chewing the given number of times.'''
    for line in range(times):
        print('I will not chew gum in class.')
```

There are two things to note here. First, the `for` loop uses a variable that we've named `line` that iterates over the list `range(times)`. For example, if `times` is 42, then `range(times)` will be the list `[0, 1, 2, ..., 41]` with 42 numbers in it. We actually don't care at all about that list, other than the fact that it has the right number of items in it. This will cause `line` to take the 42 values from 0 to 41, thereby repeating the block below it 42 times.

That block contains something new: The line `print('I will not chew gum in class.')` The `print` function prints the given string on a line to your screen, followed by a carriage return (so that the next print is on its own new line).[1] It's worth noting that `print` is different from `return` in a few ways. Most importantly for now, `print` doesn't end the execution of the function. Imagine that we replaced the line `print('I will not chew gum in class.')` with the line `return 'I will not chew gum in class.'` The first time we enter that block, Python would display that string on your screen but then the function would end. Remember, a `return` statement is powerful – it ends the function! The function as we wrote it with

---

[1] In Python 2, `print` does not require parentheses around the string to be printed, but parentheses are permitted. In Python 3, `print` is a function and thus requires parentheses. For simplicity, we always use parentheses to be compatible with both versions of Python.

print doesn't actually return anything. When it's done with the for loop, it just ends.

We still don't recommend chewing gum in class.

2.5 From Gum to CAT Boxes

Our goal is to help find genes in a DNA sequence, and we're working our way toward increasingly sophisticated ways of doing that. One tool we'll need is to locate particular sequences or *motifs* in a DNA string.

It turns out that genes are frequently "flanked" by certain motifs. One motif that appears next to many genes is called a "CAT box." It's called that, but it's actually the pattern CCAAT and it generally appears near the spot where transcription of the gene begins.

Say we've been given the task of finding and counting CAT boxes in a DNA sequence. To do this, we'll write a function called countCCAAT(DNA) that returns the number of occurrences of the CCAAT motif in the given DNA string. For example, here's what countCCAAT might look like in action:

```
>>> countCCAAT('GGCCAATGCCAAT')
2
```

Notice that input to countCCAAT is a string and we're going to need to systematically move over that string looking for CAT boxes in every position. This suggests that we should set up a for loop.

Let's start by using ideas that are familiar to us: A counter to keep track of the number of times we've seen the pattern and a for loop to iterate over the DNA:

```
def countCCAAT(DNA):
    ''' Computes number of occurrences of CCAAT in input string.'''
    counter = 0
    for nuc in DNA:
```

But now we're stuck. Do you see why? We can look at one character in the DNA string at a time as we loop over a string, but we want to look at *five* characters at a time in this case to see if we've found an occurrence of the motif 'CCAAT'.

Here's the key big idea: Instead of looping over our DNA *string*, we'll loop over the *list* of indices in that string. For example, if our DNA string is 'GGCCAATT' that string has characters at indices 0, 1, 2, 3, 4, 5, 6, 7. For example, `DNA[0]` is `'G'` and `DNA[7]` is `'T'`. The length of this string is 7, so `range(0, 8)`, or equivalently `range (8)`, will give us the list of desired indices `[0, 1, 2, 3, 4, 5, 6, 7]`. By the way, this is the reason that `range(8)` stops at 7 rather than at 8. In general, our string won't have length 7, but we can get its length using `len(DNA)`. So, our function will start like this:

```
def countCCAAT(DNA):
    counter = 0
    for index in range(len(DNA)):
```

Now what? The variable that we've named `index` can be thought of as a "finger" that tells us the position that we're currently examining. It starts off at 0. Now we'd like to see if the part of the DNA string beginning at 0 and ending at position 4 is the pattern `'CCAAT'`. If so, we'll increment the counter. If not, we're done considering that index and we let the `for` loop give us the next value. In our example with DNA `'GGCCAATT'`, the picture will look something like the one in Figure 2.4.

First, `index` is equal to 0 and we look at the slice of the string from that point 4 bases forward, that is, `DNA[0:5]` (again, remember from Chapter 0 that the 5 here means "stop at 4"). That slice `'GGCCA'` is not equal to our `'CCAAT'` motif, so we don't increment the counter within the block. Next, `index` takes the value 1 and we look at the slice `DNA[1:6]`, which is `'GCCAA'`. Again, that's not what we're looking for, so the counter isn't incremented. The next time around, index takes the value 2 and the slice `DNA[2:7]` is `'CCAAT'`. Bingo! Found one! We increment our counter.

We haven't shown what happens afterwards. Our `index` variable will next take the value 3 and the slice `DNA[3:8]` is not our pattern. Then it will take the value 4 and our slice becomes `DNA[4:9]`. That might worry you because that's asking for the characters at indices 4 through 8 and there is no character at position 8. The good news is that when it comes to slicing, Python is friendly. When it sees `DNA[4:9]`, it says "well, I'll give you as much of that slice as I can." In this case, it's `'AATT'`.

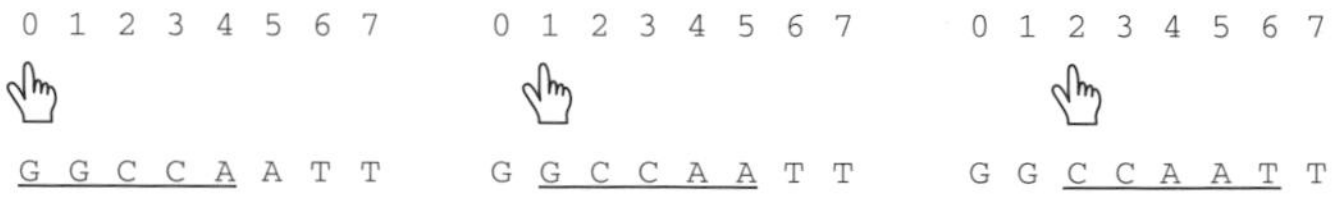

```
0 1 2 3 4 5 6 7      0 1 2 3 4 5 6 7      0 1 2 3 4 5 6 7
🖐                        🖐                        🖐
G G C C A A T T      G G C C A A T T      G G C C A A T T
```

Figure 2.4.

With slicing, Python doesn't mind that we've asked to go beyond the end of the list.

So, here's our countCCAAT function:

```
def countCCAAT ( DNA ) :
    ''' Computes number of occurrences of CCAAT in input string.'''
    counter = 0
    for index in range (len (DNA) ) :
        if DNA [index : index+5] == 'CCAAT' :
            counter = counter + 1
    return counter
```

One last thing before we let you roll up your sleeves to do some programming. Notice that our countCCAAT function has an if statement with no else. That's fine! Recall that the else is optional. In this case, if we find a match beginning at this index, we increment the counter. If not, the block is done and we return to the top of the for loop to let index take on the next value in its list.

2.6 Functions Can Call Other Functions!

Imagine that you wish to find the number of occurrences of CCAAT in a number of different DNA strings. We'd like to write a function called multiCountCCAAT that takes *multiple* DNA strings as input and prints the number of occurrences of CCAAT in each one.

We know how to write functions that take one input, or two inputs, or three inputs, etc., but how do we write a function that takes *any number* of inputs? Here's the key insight: We can write a function that takes just one input but that input is a list! That list can contain as many items as we wish. It seems sneaky, but we've consulted with our attorneys who have determined that it's entirely legal!

Our multiCountCCAAT function will take a list of DNA strings as input and will look like this when we run it:

```
>>> multiCountCCAAT ( ['GGCCAATT', 'CAT', 'CCAATCCAAT'])
1
0
2
```

We could write this function "from scratch," but notice that this function needs to do exactly what our earlier countCCAAT function does, except that it needs to do it for every string in the given list. We can, therefore, have our new multiCountCCAAT function "call" our existing countCCAAT function, once for each string in the given list. It looks like this:

```
def multiCountCCAAT( DNAList ):
    ''' Prints the number of occurrences of CCAAT in each string in
        the given DNAList.'''
    for DNA in DNAList:
        print( countCCAAT(DNA) )
```

It's that simple! The variable that we've called DNA (we could have called it anything) iterates over each of the strings in the given DNAList. For each string, we enter the block and call the countCCAAT function which returns the number of occurrences of CCAAT. We then print that number to the screen.

It was convenient to have the multiCountCCAAT call the countCCAAT function because we had already written the latter function – so why not use it!? An even more compelling reason to have one function call another function for help is that it can make your programs easier to read and understand. By dividing logical tasks into separate functions, it's generally easier to understand the logic of the program. Ultimately, it can also make finding errors or modifying the behavior of your program easier.

It's worth pointing out again the difference between return and print. In this case, countCCAAT(DNA) is being called in the line print(countCCAAT(DNA)). So, Python temporarily puts the multiCountCCAAT function "on hold," goes to countC-CAAT, passing it the current DNA string, and waits for that function to return a value. The return statement in countCCAAT causes that function to terminate and returns the value to the line containing print. At that point the value is printed (displayed on the screen) and multiCountCCAAT resumes computation.

Here's another version of the multiCountCCAAT function, called multiCountCCAAT2, that has the same behavior but does not use a helper function:

```
def multiCountCCAAT2( DNAList ):
    ''' Prints the number of occurrences of CCAAT in each string in
        the given DNAList.'''
```

```
for DNA in DNAList:
    counter = 0
    for index in range(len(DNA)):
        if DNA[index:index+5] == 'CCAAT':
            counter = counter + 1
    print(counter)
```

There are a few things to notice here. Notice that this code has a `for` loop within a `for` loop! The first `for` loop is called the *outer* loop and the second one is called the *inner* loop. Loops within loops are called *nested loops*.

In this nested loop, the outer loop is ranging over the DNA strings in the given `DNAList`. For each such string, we set the `counter` to 0 and then enter the inner loop, which ranges over a list of indices whose length is specified by the current DNA string from the outer loop. When that inner loop is done, the counter's value is equal to the number of `'CCAAT'` strings found in the current DNA string. That value is then printed and the outer loop continues by selecting the next DNA string from the `DNAList` and starting the process over again.

Notice the indentation of the `print` line in this program. If the `print` line was indented to align with the line `if DNA[index:index+5] == 'CCAAT'`, the behavior of our program would have been very different: We would print for each value of `index` in the inner loop, and that's more than we intended to print. If we indented the `print` line even more to align with `counter = counter + 1`, the behavior of the program would have been different again: Now we'd print each time the counter was incremented – again printing when we had not intended to do so.

These two versions of the `multiCountCCAAT` function are another example of different ways to do the same task. Is one better than the other? It's a matter of taste. However, the first version demonstrates the idea of modular design where one function calls another function for help. That's a powerful idea and one that is quite useful when building larger and more complex programs.

You're now ready for the task that we set out at the beginning of this chapter: A program that locates potential pathogenecity islands by locating areas of the genome whose GC content is substantially higher than the whole-genome GC content.

Putting it All Together

In this chapter we've seen that `for` loops are just as happy to loop over lists as they are to loop over strings. We've seen that functions can call other functions and we've seen the difference between `print` and `return`.

In the programming problems for this chapter, you'll first generalize the `countCCAAT` function to find the number of occurrences of any motif – not just CAT boxes. It turns out that there are other recurring motifs in DNA and counting the number of occurrences of these patterns can be helpful in determining the function of a particular region of the genome. In the second problem, you'll write a program that reports the changes in GC content as we move across a section of DNA. As we noted at the beginning of the chapter, changes in GC content are useful in identifying pathogenicity islands.

Open Reading Frames and Genes

3

> **Computing content:**
> top-down design of larger
>   programs, randomness,
> computational experiments

In the previous chapter, we saw how computational methods can be used to find pathogenicity islands in DNA. But now we want more! Specifically, we want to find actual genes in these pathogenicity islands so that we can analyze them and determine their function.

In this chapter, you'll write a program that finds candidate genes in the genome. We say "candidate" because some things that look like genes turn out not to be. In the next and final chapter of this unit, you'll take the last step to write a program that determines which of the candidate genes are likely to be real. You'll then use your gene-finding program to identify the genes in a pathogenicity island of *Salmonella typhi*. Finally, you'll use a web-based search tool called BLAST to compare these genes to known genes in the GenBank database. This final comparison will allow you to infer the function of some of the genes you've identified.

3.1 Open Reading Frames and the Central Dogma

Genes carry the instructions for proteins, which are the main molecules that "do things" in cells. To be able to find genes we must know something about how they operate. As you may recall, the process of constructing a protein proceeds according to the central dogma of molecular biology: The sequence of nucleotides in the DNA of a gene is transcribed into the sequence of nucleotides in a messenger RNA. This messenger RNA, in turn, is read off in units of three nucleotides, called

Figure 3.1. An open reading frame (ORF).

codons. Each codon specifies a particular amino acid, and the sequence of codons in the messenger RNA determines a particular sequence of amino acids in the protein.

So, the protein-coding DNA sequence consists of a series of codons. Specifically, it begins with an ATG start codon, continues through a series of other codons specifying amino acids, and ends at one of the *stop codons* TGA, TAG, or TAA. The start codon defines the proper *reading frame* for the gene, telling the cell's machinery where to begin counting off by threes. We call the region up to but not including the next in-frame stop codon an *open reading frame (ORF)*. An example is shown in Figure 3.1. We're interested in ORFs because finding them will help us find genes.

3.2 GC Content and ORFs

Let's take a short aside to explore the question of how the GC content of a DNA sequence affects the number of start codons in that sequence. After all, if you happened to know that some section of the genome had zero GC content then there couldn't possibly be any genes starting there since the start codon, ATG, contains a G!

Our objective here is to write a Python program to perform a computational experiment or *simulation* to help us determine the relationship between GC content and the number of start codons. This will be the first of several simulations that we design, program, and use to explore biological phenomena. Simulations are an important technique in modern biology.

Specifically, we'd like to write a function that takes as input a GC content value between 0 (no G's and C's) and 1 (all G's and C's), generates a random DNA sequence with that GC content, and returns a number that is the fraction of codons in the sequence that are start codons.[1]

---

[1] The truth is that it's possible to solve this particular problem analytically (mathematically) rather than computationally. However, the big ideas that we'll learn here will be useful for our gene-finding program, a case that truly requires computation.

How long should the random sequence be? It might be that we get more accurate results for longer sequences at the expense of taking more computation time. It's hard to know what value to use in advance, so in addition to GC content, we'll also let the user stipulate the length of the random DNA string in an input appropriately called `length`.

Generating just one random DNA string might not be enough. For example, if your friend gave you a trick coin that lands heads more often than tails, and you wanted to know just how unfair that coin was, you wouldn't learn much by flipping the coin just once. In general, a simulation requires that we perform the experiment many times and then average the results. For the unfair coin, flipping it 100 times and counting the fraction of outcomes that were heads would be more useful than flipping once or even a few times. So, we'll add a third input to our function, called `trials`, that lets the user specify how many times to repeat the experiment – that is, how many random DNA sequences to generate. So, let's call our function `startCodons`. With the three inputs that we've discussed here, it will begin like this:

```
def startCodons( GCcontent, length, trials ):
    ''' Given the desired GC content (between 0 and 1), the length of
        the random DNA strings, and the number of trials to perform,
        returns the average fraction of codons that are start codons.'''
```

What's next? We need a way to generate random DNA strings of the given length with the given GC content and we need a way to count the number of start codons in such a sequence. Those are two different tasks and it makes sense to write a separate function for each of them. Then, our `startCodons` function can "call" those two functions to do its job. As we've noted earlier, this "modular design" principle is an important idea when writing "large" programs – we divide up the program into smaller logical tasks and write a function for each such task. This allows us to test our functions individually before assembling them into our final program and it makes our programs much easier to read, understand, and modify later.

So, we'll need a function called `genString` that generates random strings and another called `countStarts` that computes the number of times it finds a start codon in a DNA sequence. These functions will begin like this:

```
def genString( GCcontent, length ):
    ''' Given the GC content (between 0 and 1) and the desired length,
        returns a randomly generated DNA string with the given GC
        content and length.'''


def countStarts( DNA ):
    ''' Given a string of A, T, C, and G's, returns the number of
        codons that are ATG.'''
```

It doesn't really make any difference which of these functions we write first since they are each doing their own thing. It seems to us that `countStarts` might be easier since we've seen things like it already. So, let's begin there.

3.3 The `countStarts` Function

Here's an example of what countStarts should do:

```
>>> countStarts('CCATGCATG')
2
```

There are two occurrences of `'ATG'` in our input so we get back the number 2. The fact that these ATGs are not in the same open reading frame is not important to us here – we just want to count the total number of ATGs because those are indicators of possible genes.

Since we're looking for a pattern of length greater than 1, we'll need to use the technique that we saw in the previous chapter: Looping over the indices of the DNA string and then slicing three symbols at a time to compare them with the start codon `'ATG'`.

```
def countStarts( DNA ):
    ''' Given a string of A, T, C, and G's, returns the number of
        codons that are ATG.'''
    counter = 0
    for i in range(len(DNA)):
        if DNA[i:i+3] == 'ATG':
            counter = counter + 1
    return counter
```

3.4 The `genString` Function

Now, let's work on the `genString(GCcontent, length)`. It needs to construct a random string of the given `length`. Let's call that string `output`; we'll set it initially to be the empty string: `"`. Clearly, a `for` loop is useful here and it will loop `length` times, adding a new randomly chosen DNA character to our `output` string at each iteration.

But how do we choose the characters at random? Fortunately, Python has a "module" that contains many useful functions that we can use to make random choices. In fact, this is one of many modules that Python offers. We'll explore others later for drawing figures, plotting graphs, and other purposes.

To use the `random` module, simply place the line...

```
import random
```

... at the top of your file. To get a complete list of all of the features of the library, you can either Google "python random" or you can run your file (even if the only line that you have written so far is that `import random` line) and then type `help(random)` at the Python prompt. Try both methods to see what you get.

For the sake of example, imagine that our `GCcontent` is 0.8 and the desired string length is 100. Our strategy will be to choose 100 characters at random, one at a time, so that each time we choose a new character, we choose a G or a C with probability 0.8 and an A or a T with probability 0.2. This strategy does not guarantee that we'll get exactly 80 G's and C's in any individual string. However, because we plan to generate many random strings and average the results, we will get 80 G's and C's on average. And the average is the important thing – so the strategy will work. Indeed, this is a common approach, frequently used in simulations.

Here's how our function will work. At each iteration we choose the next character by first choosing a random number between 0 and 1. If that number happens to be between 0 and 0.8, we'll choose either a G or a C with equal likelihood. If that number happens to be between 0.8 and 1, then we'll choose an A or a T with equal probability. In the first case (number is between 0 and 0.8) we must next choose between G and C with equal probability. We can do that by choosing a second random number between 0 and 1 and taking a G if that number is between 0 and 0.5 and otherwise taking a C. We can similarly use a second random number to decide between an A or a T. Of course, we won't always want to use a cutoff of 0.8. We'll write our program to use the `GCcontent` parameter as the cutoff, which allows the user to specify the desired GC content.

To get a random number between 0 and 1, we use the command `random.random()`. That name seems weird (and, perhaps, more than a bit random!) but here's the logic: The first `random` is telling Python to look at the `random` module that we imported into our program earlier. After the period, we specify which function in that random module we wish to use. It turns out that there is a function called `random()` in the random module that chooses a random number between 0 and 1.

There are other functions in that module as well. For example, `random.choice` takes a list as input and returns an element in that list chosen at random. The command `random.choice(['G', 'C'])` chooses one of `'G'` or `'C'` with equal probability. So, we have a few ways to choose random things. In `genString` our first choice is not necessarily done with equal probability, so we'll use the `random.random()` function for that. The second choice (between a G and a C or between an A and a T) is done with equal probability, so we'll use the `random.choice` function just to try it out. Here's our `genString(GCcontent, length)` function:

```
def genString( GCcontent, length ):
    ''' Given the GC content (between 0 and 1) and the desired length,
        returns a randomly generated DNA string with the given GC
        content and length.'''
    output = ''
    for i in range(length):
        myRandomNumber = random.random()
        if myRandomNumber < GCcontent:
            nuc = random.choice(['G', 'C'])
        else:
            nuc = random.choice(['A', 'T'])
        output = output + nuc
    return output
```

Finally, we're ready to write our `startCodons(GCcontent, length, trials)` function. Recall that this function takes as input the GC content, string length, and the number of random trials that we want to perform. For each trial, it calls genString(GCcontent, length) to generate a random DNA string of the desired length. Then it calls countStarts(DNA) to compute the number of start codons in that string. Finally, it totals up the number of start codons from each of these trials.

We don't want to return that total because it will be larger for long DNA strings and smaller for shorter ones. Instead, we wish to return the fraction of all possible codons that were start codons. Notice that if the DNA length had been 3, there would have been just one possible codon. If the length had been 4, there would have been two possible codons (because a codon has length 3, the codon would have had to start in the first or second position). In general, for a string with given length, there are length-2 possible codons. So, we divide the total number of start codons that we found by trials * (length-2) to account for the fact that each of our trials had length-2 codons. Here's our function:

```
def startCodons( GCcontent, length, trials ):
    ''' Given the desired GC content (between 0 and 1), the length of
        the random DNA strings, and the number of trials to perform,
        returns the average fraction of codons that are start codons.'''
    total = 0
    for r in range(trials):
        DNAseq = genString(GCcontent, length)
        total = total + countStarts(DNAseq)
    return 1.0 * total / (trials * (length-2))
```

3.5 while Loops and Population Genetics

We've seen that for loops allow us to repeat a block of code some specified number of times. But, sometimes we don't know in advance how many times we want to loop. For these cases, there's another type of construction called a while loop. We'll explore while loops in the context of a population genetics simulation.

In many cases, a gene will have a number of variants called *alleles*. These variants have slightly different DNA sequences, often resulting in different pheno-types (that is, different physical characteristics). The field of population genetics seeks to understand the various evolutionary processes that affect the frequencies of alleles in a population. Computational methods are a powerful tool for studying these phenomena.

For example, let's consider a population of haploid organisms. (Haploid organisms have just one copy of each chromosome and thus one copy of each gene. This is in contrast to diploid organisms that have two copies of each chromosome.) We'll examine a single gene, and for simplicity assume that there are just two alleles for this gene, which we'll denote by "A" and "a."

Our initial population will have some size, which we'll denote by n. Some of the "organisms" in the population will have the "a" allele and others will have the "A" allele.

From the initial population, our program will generate a population of the same size for the next generation. We'll then use that new population to generate the next one, and so forth. We'll repeat this process until we have a population in which only one of the two alleles remains. At that point, we say that the remaining allele has become *fixed*.

It may not be obvious that fixation will ever occur, but our program will allow us to verify that it does! Moreover, in the programming exercise at the end of this chapter, you'll generate plots that help visualize the impact of the population size and the allele proportions on the number of generations until fixation.

There are two parts to this problem. First, we need to be able to take a current population and construct the next generation. Then, we must repeat that process until we get to a population that has become fixed. Since there are two things going on here it makes sense to write two separate functions.

We'll call the first function `nextGen(population)` and it will take a `population` list as input where the list is simply a collection of `'a'` and `'A'` characters representing the alleles in that population. That function will build the next generation's population by initializing the new population list as an empty list, choosing a character from the `population` list at random, appending that character to the new population, and repeating until the new population's size is equal to the original population's size.

This process is well-suited for a `for` loop because we know exactly how many times to iterate the random choice process: The length of the input population list.

```
def nextGen(population):
    ''' Takes a list of alleles as input and returns a list of the
        same length representing a population in the next generation.'''
    nextPopulation = []  # start with an empty next population
    for i in range(len(population)):
        allele = random.choice(population)
        nextPopulation.append(allele)
    return nextPopulation
```

Notice the comment in our function that reads # start with an empty next population. In Python, anything following # symbol until the end of that line is interpreted as a **comment** to the reader. Comments are useful to explain things about the code to the reader. They are different from docstrings because comments are only visible in the code whereas doctrings can be viewed by the user of the function by simply typing "help(x)" where x is the name of the function.

Next, we'll write a function called popSim(population) that takes an initial population as input, calls the nextGen function to produce the next population, and repeats until one of the two alleles has become fixed.

This function is *not* well-suited for a for loop because we don't know in advance how many iterations will be needed before fixation. Therefore, we use a different kind of loop, called a while loop, that repeats its block of code as long as some condition is true, rather than a pre-determined number of times. In our case, the condition that leads us to continue looping is that neither allele has become fixed. Here's the popSim function:

```
def popSim(population):
    ''' Simulates consecutive generations, beginning with the given
        initial population, until fixation.'''
    while population.count('a') > 0 and population.count('A') > 0:
        population = nextGen(population)
    print(population)
```

Let's dissect the while loop in this example. Notice that immediately after the word while is the expression population.count('a') > 0 and population.count('A') > 0. This is a Boolean expression, that is, one that evaluates to either True or False. The expression population.count('a') > 0 returns True if the number of 'a' characters in the population list is greater than 0 and False otherwise. Similarly, the expression population.count('A') > 0 returns True if the number of 'A' characters in the population list is greater than 0 and False otherwise. Recall that placing an and between two Boolean expressions results in a new Boolean expression that is True precisely when both expressions are True. In this case, the new Boolean expression will be True if there are both some 'a' and some 'A' symbols in the population list. In other words, the expression will be True if the population has not reached fixation. And, in that case, we wish to construct the population of the next generation.

In general, a while loop works like this: Immediately after the word while is a Boolean expression followed by a colon. Python evaluates that Boolean expression. If it's True, then the indented block of code under the while loop line is evaluated. Next, Python comes back to the while loop line and evaluates the Boolean expression again. This process continues until the Boolean expression evaluates to False. At that point, Python finds the first line below the while loop block and continues computation from there.

In general, use a for loop when you know how many iterations are necessary and a while loop when you don't.

Here are a few more examples to help make sense of this new looping construct.

The first example revisits the problem of printing of the line 'I will not chew gum in class.' How many times will the gum1 function below print that line?

```
def gum1():
    ''' Apologizes for gum chewing.'''
    while True:
        print('I will not chew gum in class.')
```

It prints forever because the Boolean expression after the word while is always True! In order to get this program to stop printing, you would need to type CTRL-C on your keyboard.

The next example is more useful. We've noted that biologists often wish to find certain patterns, or "motifs," in a DNA or protein sequence. For example, earlier we wrote a function to count the number of occurrences of the CCAAT motif in a DNA string.

Instead of finding the number of occurrences, we might want to know the starting index of the first occurrence of the motif. Our goal in this example is to write a function called `find(motif, sequence)` that takes as input a "short" motif string and a "long" DNA or protein string and returns the index of the first occurrence of the motif in the sequence. We'll assume for now that the motif is known to be present in the sequence. For example:

```
>>> find('AAA', 'TCAAAGAAA')
2
```

One way to do this is with a `for` loop:

```
def find(motif, sequence):
    ''' Returns the index of the first occurrence of the motif in
        given sequence.'''
    for index in range(len(sequence)):
        if sequence[index:index + len(motif)] == motif:
            return index
```

Another way is with a `while` loop:

```
def find2(motif, sequence):
    ''' Returns the index of the first occurrence of the motif in
        given sequence. '''
    index = 0
    while sequence[index:index + len(motif)] != motif:
        index = index + 1
    return index
```

In the second function, recall that `!=` means "not equals."

Is one of these choices "better" than the other? In this case, it's debatable. On the one hand, we knew that iterating the length of the sequence would suffice, and that suggests that a `for` loop is appropriate. On the other hand, we didn't know in advance how many iterations it would take before we'd actually find the first occurrence of the motif, and thus one might argue that a `while` loop is preferable. In this case, and many cases, it's possible to use either `for` loops or `while` loops to get the job done. A good rule of thumb is to use `for` loops when you know exactly how many times you'll need to loop, use `while` loops when you don't know in advance how many times you'll need to loop, and use your personal preference when it seems that both will work equally well.

Putting it All Together

In this chapter we've examined the design and implementation of larger programs – programs that comprise multiple functions. We saw the idea of top-down design: Starting with our overall objective, we determine what helper functions we'll need to achieve that objective. In some cases, those helper functions might need their own helper functions. While our example program in this chapter used three functions, a large piece of software might involve considerably more.

In this chapter we also looked at the idea of using programs to conduct computational experiments or simulations. Some biologists call these "dry lab" experiments to contrast them with the "wet lab" experiments done at the bench. Simulations are widely used in biology and we'll use them again in the next chapter.

Finally, we saw a Python module, the `random` module. We'll use this and many other Python modules throughout this book.

In the programming problems for this chapter you'll begin by learning to use another Python module called `pyplot` that provides tools for plotting data. You'll use `pyplot` to plot the effect of GC content on the frequency of start codons. In the next problem, you'll learn how to load data from a file and then you'll use that in the last problem to locate open reading frames in DNA sequences.

Finding Genes (at last!)

4

> **Computing content:**
> gene finding

4.1 From ORFs to Genes

One might think that finding genes would be an easy task, simply requiring us to find open reading frames. Unfortunately, not every open reading frame is a gene. Consider the boxed sequence in Figure 4.1 below. The sequence starts with the start codon ATG and ends with a stop codon TAA in the same reading frame. However, this region is not an actual gene; it comes from a region of the *Salmonella* genome between two genes. Open reading frames like this are spurious, having arisen by random chance. The cell does not actually use them to make a protein.

Because not every ATG is a start codon, and not every open reading frame is a gene, the problem of gene finding is not just finding open reading frames, but also deciding which ones really represent genes.

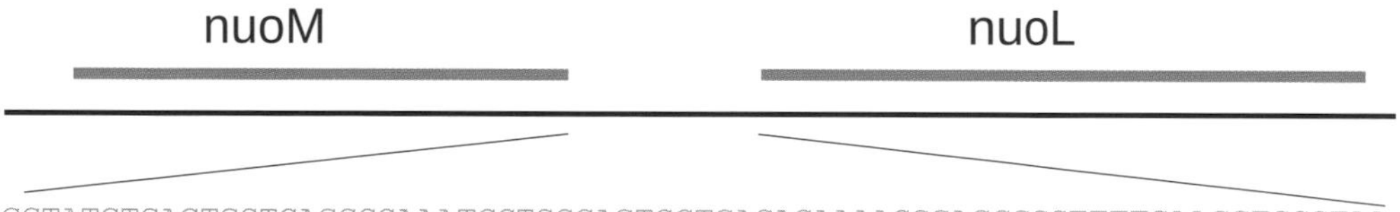

nuoM   nuoL

```
GCTATCTCACTCGTCAGCCCAAATCCTGCCAGTGCTCACACAAAACGCAGCGCGTTTTGAACGTCCGTAA
GGACGGCCCCGTAGGGGTGAGCTTCGCGAATCATCCTCACGTACTTCAGTACGCTCCGGTTGCTGTGCGC
TGGCGGTATCCGGTCTGACTTCGCCGATGACGCCTGTACTTCAGGAACAGATTTTCAACGAATTCTAAAA
ATTATTTTTGGGTTTGTAGGCCGGATAAGCACTGCGCC
```

Figure 4.1. A non-coding region in *Salmonella*, including an ORF (boxed) which is not a gene. The region is located between the nuoM and nuoL genes.

But how can we distinguish between ORFs that truly code for protein and spurious non-coding ORFs that have arisen randomly? A very simple strategy makes use of the assumption that ORFs representing true genes are significantly longer than random ORFs. Therefore, we can find likely genes by identifying ORFs that are considerably longer than ORFs arising at random.

You've already written a program to find ORFs and we'll use that function in the spirit of good modular design. To help us determine whether or not each of those ORFs is likely to be a gene, we'd like to know how long a random ORF would be. To answer that question, we'd like to compare the length of our candidate gene ORF with the longest ORF lengths in some set of comparable non-coding sequences.

In the previous chapter, we generated random DNA for our start codon experiment. We could do the same thing here, but an even easier solution – and one that guarantees that we preserve the proportion of A's, T's, C's, and G's – is to simply randomly shuffle (i.e., scramble) our DNA sequence and look for the longest ORF after each shuffle.

For example, consider the DNA sequence in Figure 4.2. We've found an ORF there which is 318 nucleotides long. To determine whether it's likely to be a gene, we shuffled the original DNA sequence and found the longest ORF in that new sequence. In our first shuffle, the longest ORF had length 126 nucleotides. In a second random

A sequence of interest

```
AATGGGCCGACCAAGGCGACATAGACGCGAATCGGACCAGACGCCGGCTCACCTGTTCATCTACCTTTCTGC
GTTGGCGCTAAAAGTTAACGATCGGGCCCTGCGCCGAAACGAAACGTCAGGAATCGACAAATACCAAGTATC       Longest ORF:
TAAGCTACGGGATAAGCCCCCCCTCGCGAGAGAGGGGAAGGGGTCAATATTTCCCTGGCCGACTGACAATGG       318 nucs
AGTGTACTTACCGGTATACAGTTTGTACTCTACAGCCATCGCTGTCTTACGACGTATTCGGGGCATTTCAAC
ATGCTGTCTCTCAGGAGTTTTCGCGCGCTGAAAGAACTCCCATCTAAACCCTG
```

Randomly shuffled versions of this sequence

```
GACAACTCTATAGTTCATGGAGCCCTGAATAGTCTATCATAGAGAAACCGGCGGGCGCCCCGGACCTATGAG
ATGAGCTTATTCGTGACTGGTTCGACCAGTAGGCAGTCGGAGCACGACTCGTGTAGCTACTTAGTTGCTTGA       Longest ORF:
AAAACCCGCGACGACATGGCCGGGGAGCCCAACTTTGAAGTGTCGCGACCTCTTAGAAATAGGTCACACACT       126 nucs
CACCTCCCTCAGCCGAGGTCCAAATCATATGGCAAGCAGCCTATCAAGGTCTCGGCCCCCAGCCGCTGGTAG
GCACCTCAAATAGCAGTTTACTTTTCCCACCAAGATGTCCGTATAGGATGGAT
```

```
CTGCCCTCGCGACGTAAAGGCCTACCCTTATTCGGGCGCTGGTGTCTGGTGTCTGCACCTTGTACGATTTAA
TCCGTCTTACGCACCGCGGGTGTCAGTGCAAAACGACTTGGGCTTACAGACATGAATCGCGGAACTCTGAAG       Longest ORF:
TATGGGTCGACCAGTCCACATTATGGAGGGAGCCAGAGTCCAACCCGGGAGGCGGGGCACCACACGCGGTAT       156 nucs
TTTAAGAGGAACCACGCTTGATCACCAACGGAAAGTAGCCGCTAAATTATCGTCAATCTACCCTCAAACACA
AAACCTCGGCTGAACGTCATATTCGAAAAGCTCTACATTTCGGGTTCAGGCCC
```

Figure 4.2. A sequence of interest along with two randomly shuffled counterparts. The longest ORF in the original sequence is much longer than the longest ORFs in the shuffled versions, which is what we would expect from a true gene.

shuffle, the longest ORF had length 156. These are both much shorter than what we found in our real sequence, suggesting the ORF in the real sequence is longer than we'd expect by chance. This leads us to believe that our original ORF of length 318 is probably a gene. Of course to be really confident we should shuffle the sequence many more times – that is, we should perform more experimental trials.

4.2 Genes Occur on Both Strands

There is one final element to consider when finding genes: They can occur on either of the two strands of DNA. Consider the region illustrated in Figure 4.3. There are two genes shown, one on each strand. For gene 1, the sequence of codons from ATG to a stop codon can be read by going 5′ to 3′ (left to right) on the top strand. For gene 2 in contrast, we find an ORF going from 5′ to 3′ (right to left) on the bottom strand. Because both of these situations are possible, when we look for genes we must search for ORFs on both strands.

4.3 Determining the Function of a Protein

Let's now return to the question of how *Salmonella* enters host cells. As we saw earlier, researchers first identified a pathogenicity island in the DNA and then identified the longest open reading frames that were present there. The next step was to determine the function of these putative genes and, ultimately, understand how *Salmonella* evolved to have them.

That's exactly what you'll do using the powerful computational tools that we've acquired over these last several chapters! In the final programming project of this unit, you'll first find ORFs in a sequence and determine whether they are genes. You'll do the second part by simulation, that is, by repeatedly shuffling the DNA, finding the longest ORF in that random sequence, and determining how the length of your original ORFs compares to the length of the longest random ORFs. You'll then use these functions to identify genes in several sequences of interest,

```
                  Gene 1
5' - AATGCCGTGC...TTGTAGACGTAGGCTTACGA...TCGTCATGGG - 3'
3' - TTACGGCACG...AACATCTGCATCCGAATGCT...AGCAGTACCC - 5'
                  Gene 2
```

Figure 4.3. Genes can be found on both strands of DNA. Gene 1 goes 5′ to 3′ (left to right) on the top strand, and gene 2 goes 5′ to 3′ (right to left) on the bottom strand.

Salmonella SpaN

```
MGDVSAVSSSGNILLPQQDEVGGLSEALKKAVEKHKTEYSGDKKDRDYGDAFVMHKETALPLLLAAWRHGAPAKSEHH
NGNVSGLHHNGKSELRIAEKLLKVTAEKSVGLISAEAKVDKSAALLSSKNRPLESVSGKKLSADLKAVESVSEVTDNA
TGISDDNIKALPGDNKAIAGEGVRKEGAPLARDVAPARMAAANTGKPEDKDHKKVKDVSQLPLQPTTIADLSQLTGGD
EKMPLAAQSKPMMTIFPTADGVKGEDSSLTYRFQRWGNDYSVNIQARQAGEFSLIPSNTQVEHRLHDQWQNGNPQRWH
LTRDDQQNPQQQQHRQQSGEEDDA
```

Shigella SpaN

```
MALDNINLNFSSDKQIEKCEKLSSIDNIDSLVLKKKRKVEIPEYSLIASNYFTIDKHFEHKHDKGEIYSGIKNAFELR
NERATYSDIPESMAIKENILIPDQDIKAREKINIGDMRGIFSYNKSGNADKNFERSHTSSVNPDNLLESDNRNGQIGL
KNHSLSIDKNIADIISLLNGSVAKSFELPVMNKNTADITPSMSLQEKSIVENDKNVFQKNSEMTYHFKQWGAGHSVSI
SVESGSFVLKPSDQFVGNKLDLILKQDAEGNYRFDSSQHNKGNKNNSTGYNEQSEEEC
```

Alignment

```
Salmonella SpaN   248 TIFPTADGVKGEDSSLTYRFQRWGNDYSVNIQARQAGEFSLIPSNTQVEH       297
                      :|......|..::|.:||.|::||..:||:..  ::|.|.|.||:..|.:
Shigella SpaN     204 SIVENDKNVFQKNSEMTYHFKQWGAGHSVSISV-ESGSFVLKPSDQFVGN       252

salmonella        298 RLH---DQWQNGNPQRWHLTRDDQQNPQQQQHRQQSGEED          334
                      :|.   .|...||  .|:..::.::.|..........:..||:
NP_085314.1       253 KLDLILKQDAEGN-YRFDSSQHNKGNKNNSTGYNEQSEEE          291
```

Figure 4.4. Sequences of the *Salmonella* and *Shigella* SpaN proteins, and an alignment between the two. In the alignment the I symbol indicates identical amino acids, and the : symbol indicates amino acids that are different, but have similar chemical properties.

including the RF319 region and a mystery sequence (which might, for example, be a part of a pathogenicity island that was discovered by examining GC content).

Once you've found likely genes this way, you'll want to understand their function. A gene's function is often similar to the functions of related genes in other species. Thus, one way to get information about a gene of interest is to identify related genes in other species. But how can we find related genes? Figure 4.4 shows the protein sequence of a *Salmonella* gene from the RF319 region. Its sequence turns out to be very similar to a gene from another bacterial species, *Shigella flexneri*. The *Shigella* gene was studied first, and was thought to be involved in pathogenesis.

If you just look at the two protein sequences in Figure 4.4, it's very hard to discern the similarity. However, we can use a computational method called *alignment* to identify the parts of each protein which are related. In Part II, we'll study the alignment method. For now, we'll use the BLAST alignment software available on the internet, which does something similar.

Alignments between *Salmonella* RF319 genes and *Shigella* genes suggested that this region was related to a region in the *Shigella* genome, and likely possessed a similar function. And thus the genes on RF319 were given the same names as the *Shigella* genes. Subsequent studies showed that RF319 is part of the 40,000 bp long pathogenicity island we mentioned before.

Putting it All Together

In this chapter we've developed a strategy for finding open reading frames (ORFs) that code for proteins. This strategy is based on the premise that protein-coding ORFs should be relatively long in comparison to random ORFs. To find them we perform a "computational experiment" to compare the length of each ORF in a gene to the length of ORFs in randomly permuted sequences. Techniques of this type – involving the randomization of biological data to determine the likelihood that an observation is meaningful – are used in many different applications in biology.

Gene finding has been an area of interest in computational biology for many years, and the approaches currently in use go well beyond the strategy we have employed here. Some techniques use additional information, such as the signals that the cell's machinery uses when it produces a protein from a gene. Others exploit subtle differences between coding and non-coding sequences, for example, the frequencies with which different codons appear. Another important class of gene finders, called evidence-based gene finders, compares some external sequence to the DNA in a genome, using this information to recognize where genes are located. The external sequence could, for example, be an RNA which was isolated by experimenters. Or it could be the sequence of the same gene from a closely related species. Gene finding is still a major area of research in computational biology and new methods are constantly being developed.

This chapter concludes Part I of the book. In this part, in addition to exploring some important computational methods in biology, you've learned how to program in Python. You've seen two important constructs: Conditional constructs (i.e., `if` statements) and loops (i.e., `for` and `while` loops). We've also discussed strategies for writing larger programs by breaking up a problem into multiple discrete tasks that can each be solved by a specific function.

In the final programming problem for Part I, you'll use the tools that we've developed to locate putative genes. Then, you'll use the web-based BLAST tool to find similar genes in other organisms and learn about their function. Ultimately, you'll make your own discoveries about the function and evolutionary origin of these pathogenic genes!

FOR FURTHER READING

Groisman EA and Ochman H. Cognate gene clusters govern invasion of host epithelial cells by *Salmonella typhimurium* and *Shigella flexneri*. *EMBO J* 1993;12(10):3779–3787.

Leavitt JW. *Typhoid Mary: Captive to the Public's Health*. Beacon Press, 1997.

Mills DM, Bajaj V and Lee CA. A 40 kb chromosomal fragment encoding *Salmonella typhimurium* invasion genes is absent from the corresponding region of the *Escherichia coli* K-12 chromosome. *Molecular Microbiology* 1997;15(4):749–759.

Wilson BA, Salyers AA, Whitt DD and Winkler ME. *Bacterial Pathogenesis: A Molecular Approach*, 3rd edn. American Society for Microbiology Press, 2011.

PART II

Sequence Alignment and Sex Determination

. . . .

What determines the sex of an individual organism? It turns out there are different answers for different species. In many vertebrates, including mammals and birds, sex is determined by chromosomes. In other groups, sex is not determined genetically but instead by environmental conditions. For example, in alligators the temperature of the egg during a several-week period determines the sex of the resulting hatchling. In other species, such as clownfish, the sex of an individual may change during the course of its lifetime.

A fundamental question in evolutionary biology is how such different sex-determination systems arose. In this part of the book we address one specific piece of this very big question: The origin of the avian and mammalian sex-determination systems. While both systems are chromosomal, they use different chromosomes. Did the two sex-determination systems evolve independently or arise from a single common ancestral system? Surprisingly, this question can be addressed using a computational analysis of the genes of mammals and birds. In this chapter, we'll develop a number of widely applicable biological and computational techniques. In the homework problems you'll then use these to explore the origins of the mammalian and avian sex-determination systems.

Mammals, including humans, have a so-called XY sex-determination system. Individuals with two copies of the X chromosome become females, and individuals with one X and one Y become males. In the avian system, where the sex chromosomes are referred to as Z and W, individuals with two copies of the Z chromosome are male, and individuals with one Z and one W are female.

Our question about the origin of these systems touches on an important concept in evolutionary biology: *Homology*. When a characteristic that is common to two species is shared by descent, we say it's *homologous*. For example, cows and humans both use milk to feed their young. This was also true of the most recent

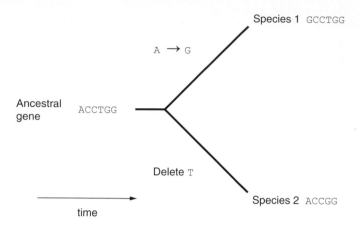

Figure 5.1. An example of homology between genes. A gene existed in the common ancestor of species 1 and 2, and was inherited in both species. We say that the gene in species 1 is an ortholog to the gene in species 2.

common ancestor of cows and humans; they inherited this characteristic from their ancestor, and thus we say milk production in cows and humans is homologous.

In this unit, we'll explore homology at the level of chromosomes. Specifically, we'd like to know whether the mammalian X chromosome is homologous to the avian Z chromosome.[1] If these two chromosomes are homologous to each other, it would mean that they both evolved from the same chromosome in the common ancestor of mammals and birds. This would suggest a certain degree of relatedness between the sex-determination systems in mammals and birds. Alternatively, the X and Z chromosomes could have evolved from different chromosomes in the common ancestor, which would imply that the sex-determination systems of the two groups evolved more independently.

Over the course of this unit, we'll develop methods to distinguish these two possibilities. Rather than looking at the entire X and Z chromosomes, we'll look just at their protein-coding regions. As we'll see in Chapter 8, determining the similarity of individual genes between the X and Z chromosomes will allow us to test homology of the entire chromosomes.

What does it mean for two genes to be homologous, and how could we recognize such a relationship? In the example shown in Figure 5.1, a particular gene was

---

[1] We focus on the Z and the X chromosomes because the W and Y are known to be degenerate copies of Z and X, respectively.

present in the common ancestor of two species and was subsequently inherited by both. Thus the two genes are homologous. When two *genes* are homologous we say that they are *orthologs*. After these two lineages diverged and began evolving independently, small changes inevitably arose in each gene. For example, in Figure 5.1, we see that the nucleotide "A" in the ancestral gene was replaced by a "G" in species 1 and the "T" in the ancestral gene was deleted in species 2. Because the changes were modest, the genes in species 1 and 2 are mostly, but not exactly, the same. We'll use the principle that if two genes are "mostly the same," they are very likely to be orthologs.

How do we measure if genes are "mostly the same?" In this unit, we'll develop a technique called the *sequence alignment score* for measuring the similarity between two genes and we'll write programs for computing those similarity scores. Ultimately, we'll use sequence alignment scores to measure the similarity between genes in a mammalian (human) X chromosome and an avian (chicken) Z chromosome. By determining the level of similarity between genes across the human and chicken sex chromosomes, we'll be able to determine which pairs of genes, if any, are orthologs. This will allow us to infer whether the mammalian X and avian Z chromosomes are homologous, and tell us something about the degree of relatedness of the two sex-determination systems.

Recursion

> **Computing content:**
> introduction to recursion

To determine whether two genes are orthologs, we need a method to measure the similarity between their DNA or protein sequences. Ultimately, we'll want a method that can handle sequences of different lengths; as we've seen, even orthologs may not have the same length due to insertions or deletions of nucleotides.

Insertions and deletions are collectively referred to as *indels*, and dealing with them is an interesting and challenging problem that we'll address in Chapter 8. For now let's assume that we want to compare two sequences of the same length, and all we need to do is determine the number of positions or "sites" at which they differ.

For example, consider the following two DNA sequences with the site indices 0 through 6 written above them.

```
0123456
ACTCATG
ATTCAGC
```

The two sequences differ at sites 1, 5, and 6; a total of three differences. We'd like to write a function called `differences(S1, S2)` that takes two string sequences S1 and S2 of the same length as input and returns the number of sites in which they differ. So, we'd use our function like this:

```
>>> differences('ACTCATG', 'ATTCAGC')
3
```

We can do this with a `for` loop without much difficulty. We first initialize a counter to 0 and then loop over the range of sites. At each site, we check whether the two sequences differ. If the strings differ at that site, we increment the counter by 1. Here's the program:

```
def differences( S1, S2 ):
    ''' Takes two strings of the same length as input and returns the
        number of positions at which they differ.'''
    counter = 0
    for site in range(len(S1)):
        if S1[site] != S2[site] : counter = counter + 1
    return counter
```

This program is perfectly good! But, next, we'll explore a very different way of approaching this problem called *recursion*. Then, we'll see that some important problems that are hard to solve with `for` loops and other methods can be solved easily with recursion. As it turns out, our task of determining the orthologs between species is a problem that requires recursion.

5.1 A Brief Diversion Before Recursion

Before introducing recursion we need to take a very short digression.

In the previous unit, we saw that it's often useful to have one function call another function for "help." For example, let's imagine that we have a function called `reverse(S)` that takes a string `S` and returns a new string that is the reversal of `S`. So, `reverse('CAT')` would return `'TAC'`. And, imagine that we have another function called `complement(DNA)` that takes a DNA string as input and returns the complementary string. So, `complement('CAT')` would return `'GTA'`. Now, we could write a function called `reverseComplement(DNA)` that takes a DNA string as input and returns its reverse complement as follows:

```
def reverseComplement( DNA ):
    ''' Takes a DNA string as input and returns the reverse complement
        string.'''
```

```
revDNA = reverse(DNA)
revComplement = complement(revDNA)
return revComplement
```

The important thing to note here is that when `reverseComplement` calls the function `reverse` for help, `reverse` does its job and then returns the value that it computed back to `reverseComplement`. At that point, `reverseComplement` resumes by assigning that value to the variable `revDNA` and then, in the next line, it calls the `complement` function for help. Similarly, `complement` does its thing and then returns its value to `reverseComplement` so that it can resume computing where it left off – in this case assigning that value to the variable `revComplement`.

This digression may not have seemed very profound to you, but here's the important observation:

Whenever one function calls another function for help, the helper function does its job and then returns control to the original function, which resumes computing where it left off.

This seemingly modest observation is at the heart of recursion.

5.2 And Now For Recursion!

Let's revisit the problem of finding the number of differences between two strings of the same length. We wrote a `differences` function earlier in this chapter using `for` loops and now we'll see another approach using recursion.

Consider two strings, `s1` and `s2`, that are equal in length. If `s1` is empty, the number of differences is 0 (since if `s1` is empty then `s2` is also empty due to our assumption that they have the same length). This kind of easy special case is called a *base case*.

Otherwise, let's look at the first symbol of `s1` and the first symbol of `s2` – that is, the symbols at site 0. If those symbols are different, we know that the total number of differences is at least one. But how many more differences are there beyond the one that we just found? That's exactly the question of how many differences there are between `s1[1:]` (the string `s1` with the 0th symbol removed or 'sliced off') and `s2[1:]`. So, we really *wish* that we had a helper function that could tell us the number of differences between `s1[1:]` and `s2[1:]`. Wishful thinking is a useful problem-solving tool – let's wish for it now and get it later!

So, imagine that we had such a function and let's call it `helper`. If `S1[0]` differs from `S2[0]`, we'd simply return `1 + helper(S1[1:], S2[1:])`. That leading 1 is to account for the difference between `S1[0]` and `S2[0]`. Next, the `helper` function is supposed to tell us how many differences there are between `S1[1:]` and `S2[1:]`.

If `S1[0]` does not match `S2[0]`, then we simply need to find the number of differences between `S1[1:]` and `S2[1:]`, which we obtain by calling `helper(S1[1:], S2[1:])` and returning that value.

So, our new function, which we'll call `differences2`, looks like this:

```
def differences2( S1, S2 ):
    ''' Takes two strings of the same length as input and returns the
        number of positions at which they differ.'''
    if S1 == ": return 0
    elif S1[0] != S2[0]: return 1 + helper(S1[1:], S2[1:])
    else: return helper(S1[1:], S2[1:])
```

But the `helper` function was the result of wishful thinking! Now we have to write that `helper` function! How will it work? It will take its own two strings as input. If those two strings differ in their first symbols, `helper` needs to return 1 + the number of differences between the two remaining strings. So, it seems that `helper` will need its own helper function! And that function will need its own helper and this quickly results in a serious headache. It doesn't look like this approach has much promise and you're probably feeling pretty skeptical about the whole wishful thinking idea. But, amazingly, we can fix this problem with almost no effort!

Notice that the `helper` function that we wished for is just trying to find the number of differences in *its* two strings. In other words, the `helper` function has the same job as the `differences2` function that it's trying to help! So, the `differences2` function can be its own helper! Here's our attempt to capture this strange idea in a function called `differences2`:

```
def differences2( S1, S2 ):
    ''' Takes two strings of the same length as input and returns the
        number of positions at which they differ.'''
    if S1 == ": return 0
    elif S1[0] != S2[0]: return 1 + differences2(S1[1:], S2[1:])
    else: return differences2(S1[1:], S2[1:])
```

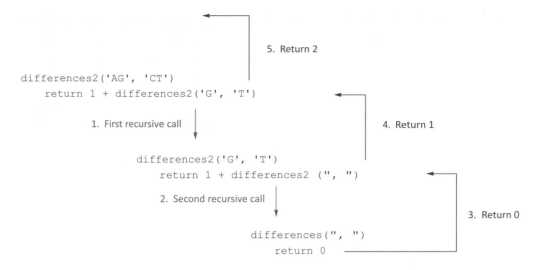

Figure 5.2. The recursive `differences2` function on inputs `'AG'` and `'CT'`.

This style of programming is called **recursion**. Recursion is characterized by functions that call themselves for help. If it seems magical, strange, or even circular reasoning, that's the normal reaction.

To get a sense of what's happening here, let's watch the `differences2` function in action when we run it on two short strings.

Consider running `differences2('AG', 'CT')`. Figure 5.2 summarizes the steps that are explained in the following paragraphs.

The function first checks whether `S1`, which is `'AT'`, is empty. It's not. So, it then checks whether `S1[0]` (which is `'A'`) is not equal to `S2[0]` (which is `'C'`). Indeed, the two symbols are different. So, `differences2` performs the line...

```
return 1 + differences2(S1[1:], S2[1:])
```

... which is effectively saying: "I will return 1 (for the difference that I just observed between `'A'` and `'C'`) plus the result of `differences2('G', 'T')`." So, off we go to find out what `differences2('G', 'T')` will return.

We're running `differences2` but now with `S1` equal to `'G'` and `S2` equal to `'T'`. We see that `S1` is not empty, so `differences2` checks whether the first two symbols in those string are unequal. Indeed they are, so `differences2('G', T')` performs the line...

```
return 1 + differences2(S1[1:], S2[1:])
```

... which says "I will return 1 (for the difference I observed between `'G'` and `'T'`) plus the result of `differences2(", ")`. I'll wait patiently to see what `differences2(", ")` gives me and then I'll take that result and add my 1 to it."

Now, `differences2` runs with `S1 = "` and `S2 = "`. It sees that `S1` is `"` and returns 0. That's the base case.

But the game isn't over yet! Recall that `differences2(", ")` returns 0 to `differences2('G', 'T')`, which was patiently waiting for an answer. Now, `differences2('G', 'T')` adds the 1 that it was holding (for the difference between `'G'` and `'T'`) to the 0 that it just received and returns this result of 1. Returns it to whom? To `differences2('AG', 'CT')`, which, you'll recall, called `differences2('G', 'T')` for help and promised to add 1 to whatever answer `differences2('G', 'T')` returned. So now, `differences2('AG', 'CT')` returns $1 + 1 = 2$ and we're done.

In this chapter, we'll delve deeper into why and how recursion works. As strange as it might seem now, in short order recursion will seem like second nature. But, first let's just look at another example of recursion in action.

5.3 The Factorial Function

In the next few short sections we'll build recursion skills that we'll need to continue our exploration of gene orthology. Our next example arises in determining the number of different proteins of a given length. For example, how many different protein sequences are there that use each of the 20 amino acids exactly once?

To construct such a sequence, we have 20 choices for the first amino acid, 19 for the next one, 18 for the one after that, and so forth, for a grand total of $20 \times 19 \times 18 \times \ldots \times 1$. In mathematics, this is denoted "20!" and pronounced "20 factorial." As we'll see in a moment, 20! is a very large number and the exclamation point notation is therefore aptly chosen.

We'd like to write a Python function to compute the factorial of any positive integer. Of course, we could do this using a `for` loop and you might want to try that on your own. But, here's a nice observation that will allow us to solve the problem very simply using recursion: The factorial of a positive integer n, defined as $n! = n \times (n-1) \times (n-2) \times \ldots \times 1$, can be expressed as $n! = n \times (n-1)!$ That's because, by definition, $(n-1)! = (n-1) \times (n-2) \times \ldots \times 1$.

So, to compute 20!, we *wish* that we had the value of 19! Then, we'd just compute 20 × 19! and we'd be done. Of course, computing 19! requires some work, but our wish can be realized by calling `factorial(19)` to get that result. And that will wish for 18! which it will obtain by calling factorial(18), and so on and so forth. Of course, we don't want to continue this forever. So we can observe that when we get to 1! the answer is simply 1 and we can stop. This is the base case.

So, here's our recursive implementation of the factorial function:

```
def factorial( n ):
    ''' Returns n! for any positive integer n.'''
    if n == 1: return 1  # Base case
    else: return n * factorial(n-1)
```

To help convince ourselves that this really works, let's do a thought experiment to see what happens when we call `factorial(3)`. When we call `factorial(3)`, the variable n is set to 3. Python checks to see whether n == 1, which it's not. So, the `else` statement is invoked and Python sees that it must return 3 * `factorial(2)`. Now, `factorial(2)` is called and n is set to 2.

You might be worried that n was set to 3 a moment ago and now it's set to 2. Will that mess things up? Fortunately, all is well. Here's the important thing to keep in mind:

Every time we call a function, it keeps its own values for its variables.

So, when we initially called `factorial(3)`, it sets n to 3 in its own private world. Then, it calls `factorial(2)` which sets n to 2 in its own private world. And, `factorial(2)` needs to return 2 * `factorial(1)` so it calls `factorial(1)` which sets n to 1 in its own private world.

But now, `factorial(1)` sees that n == 1 and so it returns 1. That's the base case! To whom does it return that value of 1? To the function that asked for it, namely `factorial(2)` which, you'll recall, wants to return 2 * `factorial(1)`. The `factorial(1)` thus is now replaced by the value 1 and `factorial(2)` computes 2 × 1 and returns that value. To whom? To `factorial(3)`, which was waiting patiently to return 3 * `factorial(2)`, which can now be evaluated as 3 × 2 = 6 and this is the value that `factorial(3)` returns. Now we're done and the value of 6 appears on your screen.

So, what is `factorial(20)`, which we sought to compute at the outset?

```
>>> factorial(20)
2432902008176640000
```

That number is a bit over 2 quintillion. If the name alone doesn't impress you, consider that a group of researchers has estimated that the total number of grains of sand on our planet is about 7 quintillion.[1] Factorials arise in many situations where we need to count the number of different possible arrangements of data. But, the important thing here was the following key idea.

Recursion is not magic. It simply follows the idea that one function can call another function for help. In recursion, we just happen to be calling *ourselves* with a smaller version of the same problem.

Don't worry, we'll do more practice and identify some strategies for writing recursive functions in the next few sections.

5.4 How to Write a Recursive Function

It takes some practice before recursion feels entirely comfortable. In this section, we'll walk through the process of designing a recursive function and then ask you to pause and try a number of exercises before moving on to the next chapter.

We've mentioned the problem of finding the reverse complement of a DNA string. This problem has two parts, reversing a string and finding the complement of a string. Let's solve each of these using recursion.

First, consider the problem of finding the complement of a string. To get started, let's write a simple function that takes a single symbol as input and returns the complement of that symbol:

```
def complementSymbol( symbol ):
    ''' Takes a string comprising a single symbol as input and returns
        its complement.'''
    if symbol == 'A': return 'T'
```

---

[1] Which Is Greater, The Number Of Sand Grains On Earth Or Stars In The Sky? http://www.npr.org/ blogs/krulwich/2012/09/17/161096233/which-is-greater-the-number-of-sand-grains-on-earth- or-stars-in-the-sky.

```
    elif symbol == 'T': return 'A'
    elif symbol == 'G': return 'C'
    elif symbol == 'C': return 'G'
```

Next, we'll use this as a helper function to compute the complement of an entire string.

When solving any problem using recursion, a good place to start is the base case. Ask yourself, "what's the smallest, simplest input that I need to consider?" Your first inclination might be to say "if the string has just one symbol then we just return the complementary symbol." That's a reasonable base case, but there's actually an even easier one: If the string is *empty* we just return an empty string.

Always try to find the easiest base case possible.

So, our recursive function will begin like this:

```
def complement ( S ):
    ''' Return the complement of a DNA string S.'''
    if S == '': return ''  # if the string is empty, return the empty string
```

Next, we need to consider the remaining case, known as the *general case*. The secret for the general case is to ask yourself the question "would it help if I could find the complement of some shorter part of the input string S?" For example, imagine that we want to find the complement of the string 'CAATG'. If we had the complement of the string without the last symbol, that *would* be helpful! In this case, the complement of 'CAAT' is 'GTTA'. Once we have that, we can add the complement of the last symbol on the end of it. In this case, the last symbol is 'G' and its complement is 'C'. So, the complement of our original string is 'GTTA' + 'C', which is 'GTTAC'.

What have we just done? We've observed that the problem of finding the complement of a string can be decomposed into the problem of finding the complement of a shorter string plus the small amount of extra work involved in finding the complement of just the last symbol. In other words, we've found a way of expressing our problem in terms of solving a smaller version of itself.

You'll notice that we used exactly this approach in the previous examples. In the case of counting the number of differences between two strings of the same length, we decomposed the problem into the smaller version of finding the number of differences between strings `S1[1:]` and `S2[1:]` and the modest extra work of adding 1 to that result if the original strings mismatched in the first position. In the case of factorial, we computed the factorial of the smaller number n − 1 and then did the extra work of multiplying that result by n.

A recursive solution is based on decomposing the problem into a smaller version of the same problem plus a small amount of extra work.

Now that we see how to compute the complement using the complement of a shorter string, we're done! We call the `complement` function to get the complement of the substring comprising all but the last symbol. (Recall that using Python's slicing notation, the substring of `S` up to but not including the last symbol is expressed as `S[0:-1]` or, more succinctly, `S[:-1]`). Then, we do a tiny bit of extra work by concatenating the symbol complementary to `S[-1]` – the last symbol of `S`.

```
def complement ( S ):
    ''' Return the complement of a DNA string S.'''
    if S == '': return ''  # if the string is empty, return the empty string
    else: return complement(S[:-1]) + complementSymbol(S[-1])
```

Before moving on, it's worth noting that instead of recursing on the substring comprising all but the *last* symbol, we could have alternatively recursed on the substring comprising all but the *first* symbol. For example, to find the complement of `'CAATG'`, we could recursively find the complement of the substring `'AATG'` and then do the extra work of concatenating the complement of the leading `'C'` to the front. Recursion would tell us that the complement of `'AATG'` is `'TTAC'`, and the complement of `'C'` is `'G'`, so our result is `'G'` + `'TTAC'`, which is `'GTTAC'`. This version of the function, called `complement2`, is shown below:

```
def complement2 ( S ):
    ''' Return the complement of a DNA string S.'''
    if S == '': return ''  # if the string is empty, return the empty string
    else: return complementSymbol(S[0]) + complement2(S[1:])
```

In both versions of the `complement` function, we solved a smaller problem and then did the extra work to complete the solution to the original problem.

Recursive functions have:

- A general case that solves a smaller version of the original problem
- A base case that stops the recursion once the problem becomes too small

5.5 Another Example: Recursive Reverse

Next, let's try to find the reversal of a string using recursion. Remember that we start with the base case. Ask yourself, "what's the simplest string to reverse?" The empty string! So, if the input is the empty string we return the empty string.

Now for the general case. Given a string like `'spam'` (normally we'd reverse DNA sequences, but our function will work just as well for canned meat products), ask yourself, "how would reversing a shorter string be helpful?" For example, we might try reversing the substring comprising all but the last symbol. In the case of `'spam'`, we'd be reversing `'spa'`, which results in `'aps'`. What we're ultimately hoping for is the reverse of `'spam'`, which is `'maps'`. Note that `'aps'` is almost what we want – we just need to put the last symbol of the input string – in this case the `'m'` in `'spam'` – in front of `'aps'` to make `'maps'`.

So, our recursive function looks like this:

```
def reverse( S ):
    ''' Return the reversal of string S.'''
    if S == '': return ''  # if the string is empty, return the empty string
    else: return S[-1] + reverse(S[:-1])
```

Alternatively, we could observe that to reverse `'spam'`, we can first reverse the substring comprising all but the first character. So, we'd reverse `'pam'` to get `'map'`. Then, we take the first character of the original string – in this case `'s'` – and concatenate it to the end of `'map'` to get `'maps'`.

Putting it All Together

One of our students once quipped, "To understand recursion, you must first understand recursion." Indeed, at this point, recursion may seem a bit like magic. For now, the main thing to keep in mind is that just as one function may call another for help with a computational task, a function can call itself for help with a task. In fact, the programming language itself doesn't distinguish between a function calling a different function and a function calling itself. In other words, recursion is not a special feature of the language, it's simply that the language says "a function can call any function it wants for help – even itself!"

With practice, recursion will become a comfortable and familiar tool. Moreover, as we'll see in the next chapter and beyond, recursion is indispensable for many challenging computational problems. Among these is the problem of measuring the similarity between genes, a task we'll need to solve to address the origins of the avian and mammalian sex-determination systems.

The Use-It-Or-Lose-It Principle

6

> **Computing content:**
> making multiple recursive calls

So far, everything we've done with recursion we could have also done with `for` loops. The recursive functions that we've seen were conceptually somewhat different from the `for` loop versions, but recursion hasn't yet given us powers to compute things that we couldn't compute before. But now it will!

6.1 Peptide Fragments

Imagine that we have a set of fragments of proteins, where each fragment has a given mass. For example, the masses of five protein fragments might be [2, 3, 8, 10, 12] in some unit of mass. In addition, we know the mass of the original protein, say 25 units for the sake of example. The question is whether or not there is a subset of our list of fragment masses that add up to the mass of the protein. In this example, the answer is "yes": fragments with masses 3, 10, and 12 add up to 25. On the other hand, if the fragments had masses [2, 15, 17, 20], we could not have found a subset that adds up to 25. For now, we're assuming that each mass in the list can be used at most once. This problem arises in the study of protein structure and has been the focus of a recent study.[1]

---

[1] Singh R and Murad W. Protein disulfide topology determination through the fusion of mass spectrometric analysis and sequence-based prediction using Dempster-Shafer theory. *BMC Bioinformatics* 2013;14(Suppl. 2):S20. doi:10.1186/1471-2105-14-S2-S20.

Our goal is to write a function called `subset(target, numberList)` that takes two inputs: a `target` number (e.g., the mass of a protein) and a list of numbers called `numberList` representing the masses of the protein fragments. The `subset` function should return the Boolean value `True` if there is a subset of the list that adds up to our target number, and `False` otherwise. So, it should behave like this:

```
>>> subset(25, [2, 3, 8, 10, 12])
True

>>> subset(25, [2, 15, 17, 20])
False
```

The order in which the numbers in the list are given should not matter, so that we would also get:

```
>>> subset(25, [10, 2, 12, 3, 8])
True

>>> subset(25, [17, 20, 2, 15])
False
```

In addition, although each occurrence of a number in the list can be used at most once, we permit the same number to have multiple occurrences. For example:

```
>>> subset(10, [2, 2, 3, 12, 3])
True
```

Notice that $10 = 2 + 2 + 3 + 3$.

Our first attempt at solving this problem might be to do something like this: Find the value in the list that is closest to the target but not larger than the target. Use that value. Now, check to see whether the remaining value can be made up with the remaining numbers in the list. For example, in the case of a target of 25 and a number list of [2, 3, 8, 10, 12], we would start by picking the 12. Now, we have to make up $25 - 12 = 13$ using numbers from the remaining list [2, 3, 8, 10]. We pick 10 and now need to make up $13 - 10 = 3$ from the list [2, 3, 8]. We see that 8 is too large, so we omit it and move on to the 3. We pick 3 and we're done! In total, we picked 12, 10, and 3 and that adds up to 25.

This kind of strategy is known as a *greedy algorithm*. It chooses what looks good at the moment without considering the effects of that decision on future decisions. While simple and tempting, greed is not always good! For example, consider the case where the target is 15 and the number list is [10, 14, 5]. The greedy approach would choose the 14 because it's the largest value less than or equal to 15. Now, we're stuck: We need to make up 15 – 14 = 1 using numbers selected from [10, 5]. But, had we not been greedy, we could have chosen the 10 and the 5 and succeeded.

Greedy algorithms often fail to give the correct answer.

So, we need a better strategy and recursion is the secret!

Recall that our function `subset(target, numberList)` takes a number and a list of numbers as input. We'll begin with the base case. We ask ourselves, "What input values would make this problem very easy to solve?" If the target number is 0, then we don't need to select anything from the list and the answer is `True`. This is indeed an easy case, and makes a good base case. But the function has two inputs, so we should ask ourselves about an easy case for the second one, the `numberList`. The simplest case for the `numberList` is when that list is empty! What should the function return in that case? If the list of numbers is empty, we can't make up the target number (unless the target is 0, but we've already handled that situation in the first base case). So, our function begins like this:

```
def subset( target, numberList ):
    ''' Returns True if there exists a subset of numberList that adds
        up to target and returns False otherwise.'''
    if target == 0: return True
    elif numberList == []: return False
```

In general, a good rule-of-thumb is that there should be one base case for each input of the function.

Next, we consider the general case – the situation when the base cases do *not* apply. It can be helpful to reason about the general case by looking at a small example. Here, let's explore what happens when we're confronted with the problem `subset(25, [30, 10, 2, 7, 8])`.

Let's look at the first number in the list. In this example, it's 30. This is an easy case: 30 is larger than 25 so we can't use it. Therefore, `subset(25, [30, 10, 2, 7, 8])` should return whatever answer `subset(25, [10, 2, 7, 8])` returns.

So, we now consider the problem `subset(25, [10, 2, 7, 8])`. Again we can operate by considering the first number in the list. Should we use that 10 or not? It's hard to know at this point, but there are only two possible scenarios: Use 10 or don't use 10. In other words, use it or lose it! The important observation here is that those are the only two possible options. Any possible solution either has a 10 in it or it doesn't. And, as we'll see, we can let recursion do the work of considering these two scenarios.

If we use 10, then we need to make up 25 − 10 = 15 from the remaining list of numbers `[2, 7, 8]`. If we lose 10, then we need to make up 25 from the remaining list of numbers `[2, 7, 8]`. So, we call `subset(15, [2, 7, 8])` and `subset(25, [2, 7, 8])`. If one **or** both of those two recursive calls returns `True`, the answer to our original problem is `True`. If both of those recursive calls return `False`, the answer to our original problem is `False`.

So, our function looks like this:

```
def subset( target, numberList ):
    ''' Returns True if there exists a subset of numberList that adds
        up to target and returns False otherwise.'''
    if target == 0: return True
    elif numberList == []: return False
    elif numberList[0] > target: return subset(number, numberList[1:])
    else:
        useIt = subset(target - numberList[0], numberList[1:])
        loseIt = subset(target, numberList[1:])
        return useIt or loseIt
```

Let's take a closer look at `subset` to understand how it works. With recursive functions it's a good idea to ask, "does it always give me the right **type** of output?" In this case, we specified that `subset` should return a Boolean – that is, it should return either `True` or `False`. The base cases clearly do this. Moreover, the general case returns a Boolean as well: the two variables `useIt` and `loseIt` are each Booleans because they were returned by the `subset` function. And, thus, the expression `useIt or loseIt` will also be a Boolean because when we take the `or` of two Booleans, we get another Boolean.

But does the statement `useIt or loseIt` give us the right answer? Recall that `or` works like this: Given an expression of the form `A or B`, where `A` and `B` are both expressions with Boolean values, the expression `A or B` evaluates to `True` if at least one of `A` or `B` is `True`. So, imagine that `useIt` is True. In that case, our answer should be `True`. And it will be since `useIt or loseIt` will be `True` regardless of the value of `loseIt`. Similarly, if `loseIt` is True, `useIt or loseIt` will be `True`, as it should be. But, if both `loseIt` and `useIt` are `False`, the expression `useIt or loseIt` will be `False`, as it should be, and we're done!

As an example, Figure 6.1 illustrates what happens when we run `subset(6, [7, 2, 5, 4])`. The top part of Figure 6.1 shows the recursion proceeding from the top down to the base cases. The bottom part shows the results returned by the base cases at the bottom and propagating upwards.

The recursion that we used here is different from the recursion that we saw in Chapter 5. The difference is that our `subset` function makes *two* recursive calls to consider *two* possible options.

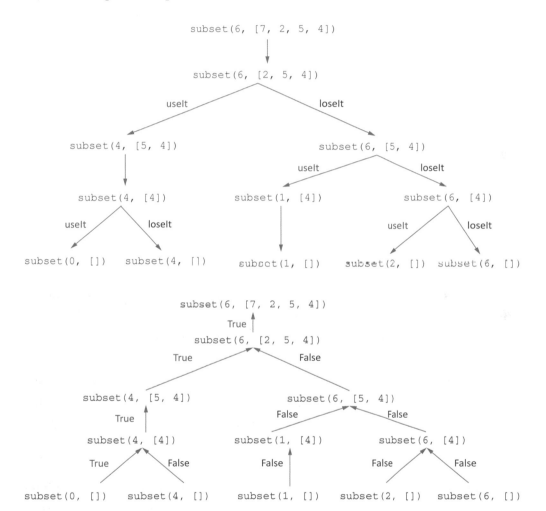

Figure 6.1. The recursive calls that take place when running `subset(6, [7, 2, 5, 4])`. The upper part shows the recursive calls from the first one at the top to the base cases at the bottom. The lower part shows the results being returned from the base cases at the bottom and propagating upwards.

Although it's not a universally accepted term, we call this *use-it-or-lose-it recursion* to remind ourselves that the recursion is exploring two options: Either we use a given piece of data or don't use it. Those are the only two options, and thus as the recursion continues it ultimately explores every possible candidate solution to the problem.

6.2 Making Change

The next problem that we consider is not biological, but it's useful in building our use-it-or-lose-it skills. The problem is that of dispensing the least number of coins to make up a given amount of change. For example, imagine that we're using a denomination system that has coins with values 1, 5, 10, 25, and 50. To make change of value 48 in this system, we'd give one coin of value 25, two of value 10, and three of value 1, using a total of six coins. That's the least number of coins required to make up a value of 48 in this denomination system.

It might seem obvious that to give out the least number of coins for a given amount of change, we should do the following: Find the coin of largest value that does not exceed the amount of change to be given. Dispense that coin and then repeat this process until we've made the requisite amount of change.

This "greedy" approach works for many denomination systems, but not all. For example, consider the "old" British coinage system before it was changed to the current system on "Decimalization Day," February 15, 1971.[2] That system had coins with values 1, 3, 6, 12, 24, 28, and 60. Imagine wanting to make 48 units of currency in this system. The greedy algorithm would first dispense a 28 unit coin and would then make up the remaining balance with a 12 unit coin, a 6 unit coin, and two 1 unit coins for a total of five coins. But we could make up 48 units of currency instead using just two 24 unit coins. Once again, we see that greed is not always good.

Our goal is to write a Python function called change that takes two inputs: The amount of change to be made and a list of the values of coins in a given denomination system. For simplicity, we'll assume that the coins in the list are given from largest to smallest value and that the last coin always has value 1 (otherwise, the denomination system does not allow us to make change for some amounts).

---

[2] http://news.bbc.co.uk/onthisday/hi/dates/stories/february/15/newsid_2543000/2543665.stm.

So, we'd like our function to work like this:

```
>>> change(48, [60, 28, 24, 12, 6, 3, 1])
2

>>> change(48, [25, 10, 5, 1])
6
```

We've seen that greedily trying to use the largest coin does not always work. Fortunately, we can use recursion to solve this problem!

We'll begin with the base case. We ask ourselves, "what input values would allow us to solve this problem easily?" Certainly, if we want to make no change, that is amount is zero, we need no coins! So, our function begins this way:

```
def change( amount, denominations ):
    ''' Returns the least number of coins required to make the given
        amount using the list of provided denominations.'''
    if amount == 0: return 0
```

Recall that when there are multiple inputs to the program, we generally need a base case for each of them. So, now we turn to the denominations list and consider the easiest case for that input. If the list is empty – that is, there are no coins available – it's certainly an easy case: We can't make change! (Unless, of course, the amount happens to be 0. But we've already addressed that base case.)

But what do we return if the denominations list is empty? Let's make a note-to-self that we'll need to figure this out later, and just write our function like this for now:

```
def change(amount, denominations):
    ''' Returns the least number of coins required to make the given
        amount using the list of provided denominations.'''
    if amount == 0: return 0  # if the amount is 0, no coins are required
    elif denominations == []: return ???  # We'll come back to this!
```

We now turn to the general case and again look at the first coin in the denominations list. If that coin's value is larger than the amount, we must "lose" that coin. If that coin's value is not larger than the amount, we'll let recursion decide whether we should use it or lose it.

For example, consider the problem change(34, [30, 20, 7, 5, 1]). We wonder whether or not to use a coin of value 30. One option is that we use that coin. In that case, we've just used one coin and we now need to know how many more coins would be necessary to make up the remaining amount, $34 - 30 = 4$, using our coin denominations. To find out how many coins are needed to make up the value 4, we need to know change(4, [30, 20, 7, 5, 1]).

On the other hand, perhaps we shouldn't use a coin of value 30. In that case, we still want to make up the value 34, but we can now remove (or "lose") coins of value 30 from future consideration. So, now we need to know the value of change (34, [20, 7, 5, 1]).

The important observation here is that there are just two options: We either use a coin of value 30 or we don't. So, 1 + change(4, [30, 20, 7, 5, 1]) will give us the least number of coins in the "use-it" case, and change(34, [20, 7, 5, 1]) gives us the least number of coins in the "lose-it" case. The lesser of these two options is the least possible number of coins needed to make up the value 34 in our denomination system.

So, our change function might look something like this:

```
def change( amount, denominations ):
    ''' Returns the least number of coins required to make the given amount
        using the list of provided denominations.'''
    if amount == 0: return 0
    elif denominations == []: return ???
    elif denominations[0] > amount: return change(amount, denominations[1:])
    else:
        useIt = 1 + change(amount - denominations[0], denominations)
        loseIt = change(amount, denominations[1:])
        return min(useIt, loseIt)
```

Notice that useIt gives us the least number of coins assuming we use the first type of coin in the denominations list and loseIt gives us the least number of coins assuming that we don't use that type of coin.

Finally, we return to the question of what should be done if the denominations list is empty. A very small thought experiment will help us here. Let's imagine what happens when we run change(1, [1]). The answer should be 1, since we can make up the value 1 using one 1 unit coin.

When we run change(1, [1]), the amount is not 0 and the denominations list is not empty, so the base cases do not apply. Moreover, the value of denominations[0] is 1, which is not greater than the amount, which is also 1. So, the `change` function looks at the general case which comprises `useIt` and `loseIt`. In this case:

```
useIt = 1 + change(0, [1])
loseIt = change(1, [])
```

In the first case, `change(0, [1])` encounters a base case because its amount is 0. So, `change(0, [1])` returns 0 and `useIt` is therefore $1 + 0 = 1$. This is the answer that we're expecting. However, `loseIt` also encounters a base case because its `denominations` list is empty. Ultimately, we'll return the lesser of `useIt` and `loseIt`.

So, if we had a base case like...

```
elif denominations == []: return 0
```

... we would have a problem, because then `loseIt` would be 0 and the minimum of `useIt` (which was 1) and `loseIt` (which is 0 now) would be 0, and that's certainly not right!

We see that in this case `loseIt` needs to be large, so that the minimum of `useIt` and `loseIt` is the value 1 found by `useIt`. In other words, if `denominations == []` then there is no solution and we should return a value so large that `useIt` will "win" every time in the line `return min(useIt, loseIt)`.

How large a value should we return if `denominations == []`? Is returning 1000, for example, a good idea? What if `useIt` is 1001? That's, admittedly, a lot of coins, but it's theoretically possible. If we took the minimum of 1001 and 1000, 1000 would "win" and we'd get back the wrong answer.

One value that is definitely sufficiently large is infinity and we can get that value in Python this way: `float('infinity')`. Try this:

```
>>> float('infinity')
inf

>>> min(42, float('infinity'))
42
```

So, finally, our change function looks like this:

```
def change( amount, denominations ):
    ''' Returns the least number of coins required to make the given
        amount using the list of provided denominations.'''
```

```
if amount == 0: return 0
elif denominations == []: return float('infinity')
elif denominations[0] > amount: return change(amount, denominations[1:])
else:
    useIt = 1 + change(amount - denominations[0], denominations)
    loseIt = change(amount, denominations[1:])
    return min(useIt, loseIt)
```

6.3 Longest Common Subsequence

This entire foray into recursion was motivated by our need to measure the similarity between two biological sequences – either DNA or protein sequences. That's the objective of this section and much of the remainder of this unit.

A simple first approach to determining the similarity between two sequences is to find the length of the longest pattern, or *subsequence*, that is common to both sequences. For example, consider the protein sequences 'DQAQ' and 'SDASQ'. The longest common subsequence here is 'DAQ'. We can find 'DAQ' as a subsequence in 'DQAQ' and also in 'SDASQ'. A subsequence is a string read left to right where skipping symbols is permitted. The reason for allowing symbols to be skipped is to capture the fact that evolution may have caused some insertions or deletions of characters in a sequence. So, 'DAQ' is said to be a longest common subsequence, or LCS, of the strings 'DQAQ' and 'SDASQ'.

For now, we'd like to have a function called LCS(S1, S2) that simply returns a number that is the *length* of a longest common subsequence of strings S1 and S2. For example:

```
>>> LCS('DQAQ', 'SDASQ')
3

>>> LCS('ASQDAQ', 'SAAQSAQ')
4
```

In the second case, there's a common subsequence of length 4: 'ASQDAQ' and 'SAAQSAQ'. There's also another different common subsequence of length 4: 'ASQDAQ', 'SAAQSAQ'. However, there is not a longer longest common subsequence, so the length of the LCS is 4.

Our goal is to write the LCS function in Python. One's first intuition might be to do something using for loops like this. Look at the first symbol in each string. If they match, we get one "point" toward the LCS "score" and we then continue by looking at the second symbols of each string, and repeat. If, however, the first symbols of the two strings don't match, we simply discard them and move on to the next symbols of each string.

Unfortunately, this greedy approach doesn't work. For example, consider the problem of finding the LCS of DNA sequences `'CAT'` and `'ATG'`. The LCS has length 2 (`'AT'` is the longest common subsequence). Using our initial idea, we'd observe that the `'C'` in `'CAT'` does not match the `'A'` in `'ATG'`, so we'd discard both symbols and look for the LCS in `'AT'` and `'TG'`. Here again the first two symbols don't match so we'd now look for the LCS of `'T'` and `'G'`. Those symbols don't match and we'd incorrectly report that the LCS of `'CAT'` and `'ATG'` has length 0.

This problem seems tricky! But here's a key observation about computing the LCS length of two strings `s1` and `s2`. First, if either `s1` or `s2` is empty, the length of the LCS is 0 since there is no subsequence in common between these two strings other than the empty string. That's a base case.

On the other hand, if `s1` and `s2` begin with the same symbol, then that symbol is part of a longest common subsequence of `s1` and `s2`, so the length of the LCS is simply 1 plus the length of the LCS of the remaining two strings, `s1[1:]` and `s2[1:]`. That can be found by recursion.

The subtle part is what happens when `s1` and `s2` do *not* begin with the same symbols, as in the case of `'CAT'` and `'ATG'`. As we saw, we can't simply remove the leading symbols from *both* strings – one of those symbols *might* be involved in a longest common subsequence with other symbols that come later in the other sequence. The trouble is that we don't know whether we should remove the leading symbol from the first string, the leading symbol from the second string, or both! In the case of `'CAT'` and `'ATG'`, the right thing to do is remove the leading `'C'` from `'CAT'`, thereby recursing on `'AT'` and `'ATG'`, which will ultimately give us the LCS of length 2. But it's hard for our program to know that in advance.

Recursion is the secret ingredient here. We simply try all *three* options and see which one gives us the best solution! The important observation is that if the leading symbols of the two strings don't match, we only have three options – lose the first symbol in `s1`, lose the first symbol in `s2`, or lose the first symbols of both `s1` and `s2`. This is an extension of the use-it-or-lose-it idea but now we're considering three possible options.

In the case of `'CAT'` and `'ATG'`, we would make one recursive call to find the LCS of `'AT'` and `'ATG'` (losing the `'C'` in `'CAT'`), another one to find the LCS of

'CAT' and 'TG' (losing the 'A' in 'ATG'), and yet another one to find find the LCS of 'AT' and 'TG' (losing the first symbols of both strings). Each of those recursive calls will give us an LCS score and the largest of those three scores will be our best solution.

So, our LCS function looks like this:

```
def LCS( S1, S2 ):
    ''' Return the length of a longest common subsequence of strings S1 and S2.'''
    if S1 == " or S2 == ": return 0 # one or both strings are empty
    elif S1[0] == S2[0]: return 1 + LCS(S1[1:], S2[1:])  # leading symbols match
    else:
        option1 = LCS(S1[1:], S2)
        option2 = LCS(S1, S2[1:])
        option3 = LCS(S1[1:], S2[1:])
        return max(option1, option2, option3)
```

Figure 6.2 shows a pictorial representation of what happens when we run LCS ('CAT', 'ATG'). At the top of the figure we see that since neither string is empty, and the leading symbols of the two strings don't match, we consider all three options in the else case. The left branch of the recursion tree, that is, option1, finds LCS('AT', 'ATG'). The middle branch of the tree, that is, option2, finds LCS ('CAT', 'TG'). The right branch, that is, option3, finds LCS('AT', 'TG'). Each of these branches are themselves solved recursively, leading each one to explore their own three options. We haven't shown all of the branches of the recursion

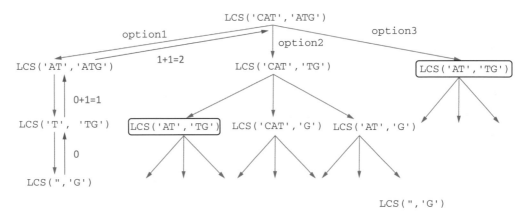

Figure 6.2. The recursive calls made by LCS('CAT', 'ATG').

because the figure would become very large! But, along the right branch, you can see how the final result of 2 is ultimately found.

Interestingly, it turns out that option3, the case where we remove the first symbol of both strings, is never necessary to consider! Our LCS function will always give us the right answer even if we omit this case. To see why, take another look at Figure 6.2. Notice that on the right side of the tree, option3 invokes LCS ('AT', 'TG'). That recursive call is in a red box. Notice that this same recursive call is also made from within option2. We've indicated that in another red box. So option3 will already be implicitly considered elsewhere and there's no need to have an explicit third option. Indeed, all of the "third branches" can be cut for the same reason, leaving us with just option1 and option2.

We can make our function even simpler by omitting that case:

```
def LCS ( S1, S2 ) :
    ''' Return the length of a longest common subsequence of strings
        S1 and S2.'''
    if S1 == " or S2 == ": return 0  # one or both strings are empty
    elif S1[0] == S2[0]: return 1 + LCS(S1[1:], S2[1:])  # leading symbols match
    else:
        option1 = LCS(S1[1:], S2)
        option2 = LCS(S1, S2[1:])
        return max(option1, option2)
```

Putting it All Together

The use-it-or-lose-it recursion paradigm that we've explored in this chapter is incredibly powerful. It allows us to write recursive functions that solve much harder problems than we were previously able to solve. Indeed, the problem of measuring the similarity between a pair of genes will require this strategy. But use-it-or-lose-it functions can become quite slow because they have to explore a huge number of different alternative solutions in search of the best one for the given problem. In the next chapter, we'll see how we make these programs fast enough to use with large pieces of data such as genes.

7 Dictionaries, Memoization, and Speed

> **Computing content:**
> the dictionary data type and its
> applications, memoization
> of recursive functions

In the previous chapter we saw how to solve some important computational problems with the use-it-or-lose-it principle. This approach obtains the correct answer by effectively exploring every possible solution to a problem. Unfortunately, it turns out that this approach can get very slow as data sets get large. For example, on a typical personal computer, running the LCS function on two random strings, each of length 10, takes approximately one thousandth of a second. But on two strings of length 25 it takes a good part of an hour and on strings of length 100 (which is still very short by the standards of biologists working with real sequences) it would take, conservatively, well over a *trillion years*.

We began this part of the book with the problem of determining homology between the mammalian X and the bird Z chromosomes. To solve this problem, we'll need to do over 1000 comparisons between proteins that are each hundreds of amino acids long. That will (almost literally) take forever!

Happily, there's an elegant and simple solution that allows us to use our use-it-or-lose-it recursive functions with very large datasets in a very reasonable amount of time. Using this so-called *memoization* method, the LCS of strings of length 100 can be found in a fraction of a second and strings of length 1000 can be evaluated in a few seconds. We think that reading this chapter will, therefore, be well worth *your* time!

7.1 Dictionaries

Before we even worry about the speed of our recursive functions, we'd like to talk about something seemingly unrelated. The connections will become evident very soon!

Biologists frequently have sequences and other data that they need to store. For example, we might want to store a set of protein sequences according to their names. A convenient mechanism for doing this in Python is something called a *dictionary.*

Here's a Python dictionary that we've called D that associates gene names (e.g., 'Gene1', 'Gene2', 'Gene3') with their corresponding protein sequences (e.g., 'MFGAA', 'MVIVY', 'LEESE').

```
>>> D = {'Gene1' : 'MFGAA', 'Gene2': 'MVIVY', 'Gene3':'LEESE'}
```

Before we explain this in more detail, here's how we can use this dictionary:

```
>>> D['Gene1']
'MFGAA'
```

```
>>> D['Gene3']
'LEESE'
```

In essence, this dictionary is allowing us to ask, "what is the sequence associated with the name 'Gene1'?" and Python responds with the answer!

Here are the rules for making a dictionary:

- The dictionary is defined using curly brackets {} at the beginning and end;
- Inside the brackets are pairs of things like 'Gene1' : 'MFGAA' where each such pair has a colon between them; and
- Those pairs are separated by commas, just like a list.

Each pair of things inside the dictionary has a first part (e.g., 'Gene1') and a second part (e.g., 'MFGAA'). The first part is called a *key* and the second part is called a *value.* Together, the pair is called a *key–value* pair.

In our example above, we named the dictionary D, but we could have given it any valid variable name that we wished. Normally, we'd use a more descriptive and useful name, perhaps GeneDictionary.

7.2 Using a Dictionary

Now that we've made a dictionary, we can use it in a number of interesting ways. As we saw earlier, one thing that we can do is this:

```
>>> D['Gene1']
'MFGAA'
```

In particular, we put the name of a key in square brackets (note the use of square brackets [] rather than curly brackets {} when *using* the dictionary) and we get the associated value.

If we give the dictionary a key that it doesn't recognize, Python won't be shy about complaining!

```
>>> D['Gene42']
Traceback (most recent call last):
  File '<stdin>', line 1, in <module>
KeyError: 'Gene42'
```

Not surprisingly, Python will be happy to tell you what keys are in the dictionary:

```
>>> D.keys()
['Gene1', 'Gene2', 'Gene3']
```

In Python 3, you'll need to do this instead:

```
>>> list(D.keys())
```

And, we can find out what values are in the dictionary:

```
>>> D.values()
['MFGAA', 'MVIVY', 'LEESE']
```

In Python 3, you'll need to this instead:

```
>>> list(D.values())
```

We can also ask whether a given key is in the dictionary:

```
>>> 'Gene1' in D
True
```

```
>>> 'Gene42' in D
False
```

Finally, after defining a dictionary, we can always add more things to it like this:

```
>>> D['Gene42'] = 'VYVYV'
```

Now, this new key-value pair will appear in the dictionary:

```
>>> D
{'Gene1': 'MFGAA', 'Gene2': 'MVIVY', 'Gene3': 'LEESE','Gene42': 'VYVYV'}
```

Notice that the dictionary was initially made using curly brackets {} and afterwards all uses of the dictionary, including adding new key-value pairs, are done with square brackets [].

Although we must make a new dictionary before using it, sometimes we don't know – or don't wish to specify – any of the key-value pairs in advance. In that case, we can start off by defining an empty dictionary and add key-value pairs to it afterwards. This is illustrated below.

```
>>> webster = {}  # We are creating an initially empty dictionary
>>> webster['chocolate'] = 'a food that contributes to happiness'
>>> webster['spam'] = 'a canned meat product'
>>> webster['chocolate']
'a food that contributes to happiness'
```

By the way, this example illustrates why Python dictionaries are called "dictionaries": Real dictionaries associate definitions with words. The words are the keys and the values are the definitions.

7.3 What Kinds of Things Can be Keys and Values?

In the examples above, the keys and values were strings. It's reasonable to wonder if dictionaries can have keys and values that are other things, like numbers or lists. Values can in fact be anything:

```
>>> D = {}
>>> D['pi'] = 3.1415926  # This value is a number
>>> D['myList'] = [1, 2, 3, 4]  # This value is a list
```

But, it turns out there are some rules about keys. A key can be a number or a string, but not a list. Here is our attempt to use the list [1, 2, 3] as a key:

```
>>> D[ [1, 2, 3] ] = 'look at this'
Traceback (most recent call last):
  File '<stdin>', line 1, in <module>
TypeError: unhashable type: 'list'
```

What's the problem with lists? We'll explain this in the next (optional) section. For now, the good news is that there's an easy workaround!

Lists have close cousins called *tuples*. A tuple is virtually exactly like a list except that it uses parentheses rather than brackets. So, while [1, 2, 3, 4] is a list, (1, 2, 3, 4) is a tuple. Tuples work just like lists in terms of indexing and slicing. Here's an example:

```
>>> myTuple = (1, 2, 3, 4)
>>> myTuple[0]
1
>>> myTuple[-1]
4
>>> myTuple[1:]
(2, 3, 4)
>>> myTuple[:-1]
(1, 2, 3)
```

The difference between tuples and lists is that tuples are not mutable. So, tuples don't support the append method that we saw in Chapter 5. Here's an example of using append with lists and our attempt to use them with tuples:

```
>>> myList = [1, 2, 3, 4]
>>> myList.append(5)
>>> myList
[1, 2, 3, 4, 5]
>>> myTuple = (1, 2, 3, 4)
>>> myTuple.append(5)
Traceback (most recent call last):
  File '<stdin>', line 1, in <module>
AttributeError: 'tuple' object has no attribute 'append'
```

In many cases, the `append` feature isn't needed, so using tuples rather than lists isn't a problem.

Finally, in the next section we'll do strange things like this:

```
>>> D = {}
>>> D[(1, (2, 3))] = 'hello'
```

Notice that the key here is the tuple `(1, (2, 3))`. That tuple has two elements, the number `1` and the tuple `(2, 3)`. On the other hand, look at what happens here:

```
>>> D[(1, [2, 3])] = 'hello'
Traceback (most recent call last):
  File '<stdin>', line 1, in <module>
TypeError: unhashable type: 'list'
```

The key we tried to use here was the tuple `(1, [2, 3])` but that tuple has a list inside it. Python reacts badly to *any* attempt to have a list *anywhere* inside a key. The takeaway message here is this:

In a Python dictionary, the key may not contain any lists.

We'll soon see the *value* of knowing this *key* piece of information.

7.4 Optional Section: How Dictionaries Work (and why you should care)

Dictionaries allow us to make key-value associations and look up the value associated with a key. In this section, we explain how dictionaries work, why they are fast, and why they don't permit keys to be mutable objects. Three great mysteries are divulged in one short section!

Imagine an actual dictionary, such as the Merriam-Webster Dictionary, where the words weren't sorted alphabetically. It would be very painful to find the definition of a word. By organizing the words alphabetically, we can search for words much faster, for example, by going to the middle of the dictionary, checking to see how our word compares to the one in the middle, and then deciding whether to continue our search in the first half or the last half of the dictionary.

This approach is great, but imagine that you need to add new words to your dictionary from time to time. (You may have been taught not to write in books, so we'll just imagine this!) Adding a new word would take a bit of effort because we'd first need to find where it belongs in alphabetical order and then write the definition at that spot.

Computer scientists developed a faster approach for this problem. It's called *hashing* and it works like this. Imagine that the dictionary contains some number of initially empty pages – say, for example, 1000 pages. Imagine also that you're very fast at finding any particular page number. (Indeed, this is something that computers are designed to do essentially instantly.) Now, for each word that we wish to add to the dictionary, we transform that word into a corresponding number. How? We might associate each letter with a number and then add up the numbers corresponding to the letters. For example, if we assign the letter "a" the number 1, the letter "b" the number 2, etc. then the word "abba" would be associated with the number $1 + 2 + 2 + 1 = 6$. We say that "abba" *hashes* to 6 in our scheme.

Now, when we want to insert the word "abba" and its definition into the dictionary, we'd go to page 6 and write the word and definition there. Later, if we wish to look up the definition of "abba," we do the conversion of letters to numbers, find that the number for "abba" is 6, go to page 6, and look up the definition.

You might be imagining all kinds of problems with this scheme. For example, there might be multiple different words that all hash to the value 6. For example, "baba" also hashes to 6 as does "ada" (since "a" has the value 1 and "d" has the value 4). This is a problem, but a modest one. If our hashing process is reasonable, not too many words will hash to the same number and we'll have at most a few words per page. So, when we hash a word and look it up on its page, we should only have to look at a few words before finding the one of interest to us.

Another problem is that some words might hash to values larger than our largest page number. Some very long word might hash to a number greater than 1000 in our scheme, which is a problem if our dictionary only has 1000 pages. Conversely, most of the words might hash to numbers well under 1000, meaning that most pages in our dictionary will go unused.

These problems too can be fixed by using more sophisticated hash functions than the one we suggested. In fact, a good deal of research has been done to design hash functions that distribute words very evenly across the available pages and Python dictionaries use very good hash functions.

Here's what Python does when you use a dictionary. When you add a key-value pair to a dictionary, Python hashes the key. It gets a number from the hash and it uses that number as the page in its metaphorical book to place the key-value pair. In Python's case, the "book" is the computer's memory, which typically has millions or billions of "pages" or *memory locations*. Python is able to access any given memory location very quickly, irrespective of the how big the page number might be.

So, when we add a word like `'baba'` into the dictionary, Python hashes `'baba'` and gets a number. That number is then used as a memory location and Python stores the key-value pair at that location. You can actually see the hash values that Python uses via the built-in `hash` function. Here's an example:

```
>>> hash('baba')
1318948168398674240
```

In truth, that hash value is much larger than the largest memory address in a computer, so your computer has to transform that number yet again to turn it into a valid memory location. But, the takeaway message here is that the hashing technique allows us to convert keys into numbers and then use those numbers as a place to store the key-value pair. And, later, to look up the value associated with `'baba'`, Python hashes `'baba'` and looks for the value in the corresponding memory location. Because the computer can access memory locations very quickly, hashing provides a very fast way to access the values associated with keys.

We can also hash numbers and tuples:

```
>>> hash(42.123)
1705667547

>>> hash( (1, 2, 3) )
2528502973977326415
```

But, we can't hash lists:

```
>>> hash( [1, 2, 3] )
Traceback (most recent call last):
  File "<stdin>", line 1, in <module>
TypeError: unhashable type: 'list'
```

Why not? Imagine that Python allowed us to hash lists and thus use them as dictionary keys. Here's what we could do:

```
>>> myDictionary = {}
>>> L = [1, 2, 3]
>>> myDictionary[L] = 'This is my favorite list'
```

In this case, L is a name for the list [1, 2, 3] which serves as a key in our dictionary. When Python hashes, it uses the list [1, 2, 3] as the input to the hash function. But now imagine that we did this:

```
>>> L.append(4)
>>> L
[1, 2, 3, 4]
```

What should happen when we ask Python for the value associated with the key referred to by L?

```
>>> myDictionary[L]
```

Python would look at L, find that it is the list [1, 2, 3, 4], and obtain the hash value for that list to find the value associated with it. The problem is that the value associated with this list was stored using the hash value for [1, 2, 3]. But now the list has changed to [1, 2, 3, 4] and that will almost certainly have a different hash value, so we'll end up looking at a memory location that is not where we stored the value!

The moral of this story is that if a key can change its value, we could store the key-value pair in one place and then look for it in a different (wrong) place later. For this reason...

Python does not permit mutable types (e.g., lists) to be the keys for dictionaries.

By the way, you're probably familiar with the term "hash" in the context of hash tags and may be wondering if there's a connection between hashing and hash tags. A hash tag is a key that allows multiple messages or tweets to be located in the same "place" (where "place" here might mean a web page or other grouping of content). Imagine that all messages about this book are tagged with a special string like #CFB. That string can then be hashed into a number and all messages with that tag will get the same number. That number, then, can be used as a location where the messages can be found.

7.5 Memoization

We now want to show you a surprising and clever use of dictionaries that we'll use to explore the similarity between genes in different species.

Use-it-or-lose-it recursion is slow. For example, let's reconsider our `change` function from Chapter 6, copied below for your reference.

```
def change( amount, denominations ):
    ''' Returns the least number of coins required to make the given
        amount using the list of provided denominations.'''
    if amount == 0: return 0
    elif denominations == []: return float('infinity')
    elif denominations[0] > amount: return change(amount, denominations[1:])
    else:
        useIt = 1 + change(amount - denominations[0], denominations)
        loseIt = change(amount, denominations[1:])
        return min(useIt, loseIt)
```

Let's run it now with `amount` 10 and `denominations` [4, 2, 1]. Figure 7.1 shows how the recursion begins.

Since the function explores two options in the general case, the recursion looks like a tree that potentially doubles in size at each level. At the first level we have the original call to the function. At the next level we have two recursive calls, then four, and so forth. If this were to continue to 10 levels, we'd have $2^{10} = 1024$

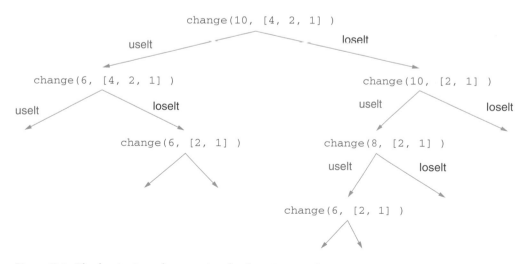

Figure 7.1. The beginning of a recursion for the `change` function.

recursive calls at level 10. If it continued to 20 levels, we'd have 2^{20} recursive calls at level 20 – and that's over a million. At 300 levels we'd have a number of recursive calls that exceeds the number of particles in the known universe. Computers are fast, but even the fastest computer is no match for exponential growth.

A close look at Figure 7.1 shows another interesting thing. Notice that the recursive call change(6, [2, 1]) appears twice in the tree, once on the left side and once on the right side. Each of those two recursive calls makes its own recursive calls and thus there's a lot of replicated work. In fact, in a large enough tree of this type, we could find the same thing being done many, many times. That repeated work is very inefficient.

Our solution to this problem is to keep a "memo" of the recursive calls that we've performed and the answers that they found. For example, after we compute change (6, [2, 1]) on the left side of the tree in Figure 7.1, we'll store the result in our memo. The result in this case will be 3, since the value 6 can be made up of three coins of value 2. Then, later, when we go to compute change(6, [2, 1]) on the right side of the tree, we'll first check in our memo to see if we've already solved that problem. Since we have indeed already solved that problem, we simply get the value from our memo and short-circuit all of the work that the recursion would have done!

It's a nice idea, but how does this memo work *in practice*? Dictionaries! A dictionary stores key-value pairs. In our case, a key will be a specific set of inputs for the change problem – that is, an amount and a denominations list. The associated value will be the solution to that problem. For example, for the problem 6, [2, 1] the associated value will be 3.

But how do we make 6 and [2, 1] into a key? That's a number *and* a list. We've seen already that a key must be a *single* item and that it cannot contain lists anywhere. The solution is to represent the problem of having an amount of 6 and list of [2, 1] using a single tuple (6, (2, 1)). In general, our memo dictionary will have keys of the form (amount, denominations) where amount is a number and denominations is a tuple of coin types. We'll simply use tuples instead of lists in order to keep our memo dictionary happy.

So, we'll memoize using the following generic approach:

- We add to our function an extra input: a memo dictionary. When we first call the function, we'll pass in an empty dictionary, but the dictionary will get populated with solutions from recursive calls over time.

- The first thing we do in our function is check for base cases.
- If the base cases don't apply, we check to see if the solution to our problem is in the memo. If so, we return that solution and we're done.
- If the solution is not in the memo, we let recursion find the solution but then insert that solution into the memo for future reference.

Using this approach, our memoized change function looks like this, where we highlight in red all of the places where we've changed the function from the previous version.

```
def memoizedChange( amount, denominations, memo ):
    ''' Returns the least number of coins required to make the given
        amount using the tuple of provided denominations.'''
    if amount == 0: return 0
    elif denominations == (): return float('infinity')
    # Since the base cases don't apply, check the memo to see if we've
    # solved this problem before.
    elif (amount, denominations) in memo: return memo[(amount, denominations)]
    elif denominations[0] > amount:
        solution = memoizedChange(amount, denominations[1:], memo)
        memo[(amount, denominations)] = solution
        return solution
    else:
        useIt = 1 + memoizedChange(amount - denominations[0], denominations, memo)
        loseIt = memoizedChange(amount, denominations[1:], memo)
        solution = min(useIt, loseIt)
        memo[(amount, denominations)] = solution
        return solution
```

Let's consider what happens when we run `memoizedChange` like this:

```
>>> memoizedChange(10, (4, 2, 1), {})
```

Notice that the second input is a tuple rather than a list; we've decided to use tuples exclusively because they work as dictionary keys. The third input to the change function is a dictionary – our memo – and it starts off empty.

When we first invoke this function, we check to see if this is one of the two base cases. It's not, so we come to the line:

```
elif (amount, denominations) in memo: return memo[(amount, denominations)]
```

This is asking if we've seen this particular problem before: An amount of 10 and a denominations tuple, (4, 2, 1). We check whether (10, (4, 2, 1)) is in the memo. The memo is empty, so the answer is "no, we haven't seen this before."

Next, we reach the line...

```
elif denominations[0] > amount:
```

... which doesn't apply in this example since amount is 10 and denominations [0] is 4.

So, we enter the body of the function and ultimately perform two recursive calls:

```
useIt = 1 + memoizedChange(10 - 4, (4, 2, 1), memo)
loseIt = memoizedChange(10, (2, 1), memo)
```

The useIt case must be completed before the loseIt case can begin. And, recall from Figure 7.1 that in the process of computing the useIt case we'll compute the solution in the case where amount is 6 and denominations is (2, 1). (Figure 7.1 shows the calls to the original non-memoized function change. But the tree of recursive calls for memoizedChange will have the same structure.) What happens when we compute memoizedChange(6, (2, 1), memo), where amount is 6 and denominations is (2, 1)? The function checks whether (6, (2, 1)) is in the memo, which it's not. So, we do all of the work required to compute the solution. The value we come up with is 3, and we put it in the variable called solution. Just before returning solution we see the line

```
memo[(amount, denominations)] = solution
```

or, specifically in this case,

```
memo[(6, (2, 1))] = 3
```

The result that we just calculated gets stored in the memo!

Now, imagine what happens when we later pursue the loseIt side, that is:

```
loseIt = memoizedChange(10, (2, 1), memo)
```

The memo dictionary now includes all of the problems that it solved on the `useIt` side and, in particular, it contains a memo entry for `(6, (2, 1))`. So, when it embarks on the recursive call `memoizedChange(6, (2, 1), memo)` on the right side of the tree in Figure 7.1, it reaches the line:

```
elif (amount, denominations) in memo: return memo[(amount, denominations)]
```

or, in this specific case,

```
elif (6, (2, 1)) in memo: return memo[(6, (2, 1)]
```

Since the key `(6, (2, 1))` is indeed in the `memo` dictionary now, it returns the value `memo[(6, (2, 1))]`, which is `3`. All of the work that would have transpired in computing the solution from scratch is avoided! And, of course, since the memo contains many other solved problems, other recursive calls are likely to be avoided as well.

This new memoized function is very fast. We encourage you to try it out with a large amount and a large assortment of different denominations.

7.6 Memoizing LCS

Just as we memoized the `change` function, let's now memoize the `LCS` function for computing longest common subsequences. Recall that our `LCS` function looks like this:

```
def LCS( S1, S2 ):
    ''' Return the length of a longest common subsequence of strings
        S1 and S2.'''
    if S1 == '' or S2 == '': return 0  # one or both strings are empty
    elif S1[0] == S2[0]: return 1 + LCS(S1[1:], S2[1:])  # leading symbols match
    else:
        option1 = LCS(S1[1:], S2)
        option2 = LCS(S1, S2[1:])
        return max(option1, option2)
```

Here again, the recursion does redundant work as shown in Figure 7.2. In this example, we're trying to find `LCS('AAT', 'AAA')` and we see two recursive calls are created to compute `LCS('AT', 'AA')`. We'd like to use memoization here to

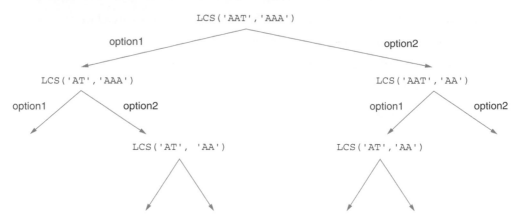

Figure 7.2. The beginning of a recursion tree for the LCS function.

store the answers that we've computed in a memo and use those answers rather than re-doing the same work twice.

In this case, a LCS problem comprises a pair of input strings, for example, 'AAT' and 'AAA'. So, our memo will have entries – that is, keys – that are pairs of strings. But, of course, a key must be a single item. We can make two items into one item by placing them into a tuple: While 'AAT' and 'AAA' are two strings, ('AAT', 'AAA') is a single tuple! So, the keys in our memo will be tuples like ('AAT', 'AAA') or ('AT', 'AA').

We'll follow the same pattern to memoize LCS as we did to memoize change: We add a memo dictionary as an additional input to our LCS function, we check to see whether the solution is in the memo before embarking on recursion, and, if not, we recurse and store the solution in the memo before returning that solution.

We've highlighted in red those lines that were added to LCS to get the memoized version:

```
def memoizedLCS( S1, S2, memo ):
    ''' Return the length of a longest common subsequence of strings
        S1 and S2.'''
    if S1 == '' or S2 == '': return 0  # one or both strings are empty
    elif (S1, S2) in memo: return memo[(S1, S2)]
    elif S1[0] == S2[0]:
        solution = 1 + memoizedLCS(S1[1:], S2[1:], memo)
        memo[(S1, S2)] = solution
        return solution
```

```
    else:
        option1 = memoizedLCS(S1[1:], S2, memo)
        option2 = memoizedLCS(S1, S2[1:], memo)
        solution = max(option1, option2)
        memo[(S1, S2)] = solution
        return solution
```

Memoization is a technique for making recursive functions fast by avoiding repeated work. A dictionary is used to store key-value pairs where the keys are the inputs to the problem and the values are the solutions that we've found. Before embarking on a recursive search, our function first checks to see whether the problem has been previously solved and, if so, it simply gets the value. If the problem has not been previously solved, the function performs all of the recursive work but then stores the solution to this problem in the memo for future use.

There's another closely related technique called *dynamic programming* that effectively does the same thing as memoization by storing the results of previously solved problems in a matrix rather than a dictionary. Dynamic programming begins at the "bottom" of the recursion tree, solving small problems first and then using those solutions to solve bigger problems higher up in the tree. We mention this only because you might hear the term "dynamic programming" in the future and we'd like you to know that it's closely related to what we've just done here!

Putting it All Together

Dictionaries provide a powerful tool for storing and retrieving data. Their structure is analogous to a book-form dictionary storing the definitions of words: For each key (word) there is an associated value (definition of the word). In programming languages like Python, the keys and values need not be words and sentences, they can be other types of data.

While we'll see that dictionaries are useful in many applications, they are particularly useful in "memoizing" recursive functions. They make it possible to store the results of problems that have already been solved, and retrieve them quickly when they are needed again.

Now that we have the use-it-or-lose-it paradigm and memoization, we're ready to return to the problem of exploring the evolution of sex chromosomes.

8

Sequence Alignment and the Evolution of Sex Chromosomes

> **Computing content:**
>
> sequence alignment

We now return to the question that we posed at the beginning of this part: Are the human X and the chicken Z chromosomes homologous? That is, did they descend from a common ancestor? To answer this question, we'll compare genes on a number of mammalian and avian chromosomes including the X and the Z. We'll measure the similarity between pairs of mammalian and avian genes and use our results to identify orthologous pairs. Finally, having found pairs of orthologs, we'll be able to assess the relationship between entire mammalian and avian chromosomes.

This approach requires a biologically reasonable way of scoring the similarity between pairs of genes. In Chapter 5, we developed the `differences` function and in Chapters 6 and 7 we developed the longest common subsequence (LCS) method. While both of these approaches provide some measure of similarity, they fail to capture some important biological processes. Thus, our first task is to develop a more realistic scoring method. With that new method in hand, we'll return to exploring the homology of sex chromosomes.

8.1 The Sequence Alignment Score

We're going to need to measure the similarity between two sequences of potentially unequal length. We'll denote these two sequences S1 and S2. To begin we'll assume that these sequences are orthologs and have descended from a common ancestor. A biologically motivated way to measure the similarity between them is to estimate the number of mutational events from the common ancestor to each of S1 and S2.

The problem though is that we generally don't know the sequence of the common ancestor – it has long since gone extinct. Therefore, we typically follow an alternative and simpler strategy. Instead of asking how many mutational events it takes to get from the common ancestor to each of S1 and S2, we ask how many events it takes to go directly from S1 to S2 or from S2 to S1. This simplification is widely used and has proven to be very effective.

In Chapter 5, we developed a simple `differences` function to measure the similarity of two sequences. The shortcoming of this function is that it requires the two sequences to be of the same length and it only compares symbols in the same positions or "sites." For example, the strings "DAQ" and "MDAPQ" cannot be compared, even though they are clearly almost identical. And, the strings "DAQS" and "MDAQ" differ at every site, even though they share the substring "DAQ" and are thus, intuitively, quite similar.

In Chapters 6 and 7, we developed the longest common subsequence (LCS) score, which seeks to maximize the number of matches between two strings. This approach behaves more reasonably with the examples above. "DAQ" and "MDAPQ" have an LCS score of 3 because they share the substrings "DAQ" and, similarly "DAQS" and "MDAQ" also have an LCS score of 3. However the LCS score is not ideal either, and can give odd results because it simply maximizes the number of matches, but doesn't penalize mismatches or differences in length. Intuitively, it seems like the strings "DAQ" and "SDASQMMMMMMMM" are less similar to each other than the strings in our cases above. But they too have an LCS score of 3! We'd expect a good measure of string similarity to reward matches and penalize both mismatches and differences in length.

Earlier in this part, we noted that differences in length can be due to insertions and deletions of nucleotides, collectively known as *indels*. Indels as well as mismatches (errors where one nucleotide has replaced another) arise due to errors in various cellular processes, for example, in DNA replication. Both should be part of a biologically sound scoring system.

The measure that we'll use in this chapter is called the *sequence alignment score* and it accounts for indels and mismatches. While sequences can also evolve in ways other than indels and mismatches, these simple mutational operations provide a reasonable first approximation to sequence evolution.

Our objective is to find the "best" way to transform one of our strings, s1, into the other, s2. For now, we'll award 1 "point" for each symbol in s1 that we leave unchanged but penalize 1 point for each symbol in s1 that we change, insert, or delete. Our goal then is to find a sequence of operations that takes us from s1 to s2 with the maximum total number of points.

In biological applications we measure the sequence alignment score for DNA or protein sequences. But for the sake of example, let's use two English words: plum and lemon. Here's one way to transform a plum to a lemon, working from left to right:

plum -> lum   (delete the p, 1 point penalty)
                   (keep the l, 1 point reward)
lum -> lem   (change the u to an e, 1 point penalty)
                   (keep the m, 1 point reward)
lem -> lemo   (insert an o, 1 point penalty)
lemo -> lemon   (insert an n, 1 point penalty)

This transformation kept the l and m unchanged for 2 reward points but carried out 4 mutation operations (1 change and 3 indels) incurring a penalty of 4. Thus, the score is computed as $2 - 4 = -2$.

Here's another possible transformation:

plum -> llum   (change the p to an l, 1 point penalty)
llum -> leum   (change the second l to an e, 1 point penalty)
leum -> lem   (delete the u, 1 point penalty)
                   (keep the m, 1 point reward)
lem -> lemo   (insert an o, 1 point penalty)
lemo -> lemon   (insert an n, 1 point penalty)

In this case, only the m in plum was left unchanged, providing a reward of 1, and the 5 mutation operations (2 changes and 3 indels) incurred a penalty of 5, for a total score of $1 - 5 = -4$. This is not as good as the previous transformation. In fact, −2 is the best possible score for transforming plum to lemon given the set of penalties we're using.

It's worth noting that we can transform `lemon` to `plum` with the same score as the transformation from `plum` to `lemon`. This is because each operation that we perform to get from `plum` to `lemon` from left to right has a corresponding "opposite operation" in a transformation from `lemon` to `plum` from right to left. This is illustrated below where the left column shows a transformation from `plum` to `lemon` and the right column shows the corresponding transformation from `lemon` to `plum`.

1. plum -> lum   (delete the p)   6. lemon -> lemo   (delete the n)
2.   (keep the l)   5. lemo -> lem   (delete the o)
3. lum -> lem   (change the u to an e) 4.   (keep the m)
4.   (keep the m)   3. lem -> lum   (change the e to a u)
5. lem -> lemo   (insert an o)   2.   (keep the l)
6. lemo -> lemon (insert an n)   1. lum -> plum   (insert the p)

In general, the best score for transforming a string `S1` into a string `S2` is the same as the best score to transform `S2` to `S1`.

Our objective is to write a function called `alignScore(S1, S2)` to compute the sequence alignment score for its input strings `S1` and `S2`. This may seem hard because there are so many different possible ways to transform `S1` to `S2`. However, recursion is ideally suited for exploring the options.

We begin with the base cases. There are two inputs, so we expect two base cases. What is the easiest possible value for `S1`? The empty string! If `S1` is empty, then we have no choice but to insert symbols one-by-one until the empty string is transformed to `S2`. This requires `len(S2)` insertions, each one incurring a penalty of 1, so the score will be `-1 * len(S2)`.

Analogously, if `S2` is empty, we must delete all the characters of `S1` for a total score of `-1 * len(S1)`. So, our program begins like this:

```
def alignScore( S1, S2 ):
    ''' Returns the sequence alignment score for S1 and S2. '''
    if S1 == '': return -1 * len(S2)
    elif S2 == '': return -1 * len(S1)
```

What happens if both strings are empty? Do we need to add another test for that? Before we do, let's check what happens with the function that we've written so far.

If the two strings are empty, the score should be 0. The first `if` statement will apply since S1 is empty and the function will return `-1 * len(S2)`, which is 0. So, there's no need here to do anything extra for the case where both strings are empty.

Next, we consider the general case that neither of the input strings is empty. In order to transform string S1 to string S2, we have several options: We can keep the first character `S1[0]` unchanged, we can change it, we can delete it, or we can insert a new character in front of `S1[0]`. Let's consider each of those options separately.

If `S1[0]` matches `S2[0]`, then we keep `S1[0]` unchanged. In that case, we get a reward of 1 point and now we can remove `S1[0]` and `S2[0]` from further consideration. In that case, we need to determine the maximum number of points that can be obtained in transforming `S1[1:]` to `S2[1:]`. Recursion can answer that for us! Therefore, our first option is:

```
1 + alignScore(S1[1:], S2[1:])
```

On the other hand, if `S1[0]` and `S2[0]` don't match, we have the option of changing `S1[0]` to `S2[0]` for a penalty of −1. Then, we can remove the first symbols from S1 and S2 from further consideration and find the largest number of points possible in transforming `S1[1:]` to `S2[1:]`. So, the second option is:

```
-1 + alignScore(S1[1:], S2[1:])
```

It's worth noting here that only one of these two options can apply, depending on whether `S1[0]` and `S2[0]` match or mismatch. So, we can more conveniently consolidate these two options into a single option as follows:

```
if S1[0] == S2[0]:
    option1 = 1 + alignScore(S1[1:], S2[1:])
else:  option1 = -1 + alignScore(S1[1:], S2[1:])
```

Next, we must consider the case that `S1[0]` is deleted. In that case, we pay a penalty of 1. Now S1 is missing its first symbol so we must find the best score for transforming `S1[1:]` to S2. So, this option will be:

```
option2 = -1 + alignscore(S1[1:], S2)
```

Finally, we can insert a new character at the front of S1. But what character should we insert in that case? Given that our goal is to convert S1 to S2, it's not difficult to see that the best choice is to insert a copy of S2[0] at the front of S1 so that those sites match and we can move on to the remaining parts of those strings. After paying a penalty of 1 for this insertion, we are now confronted with finding the best score for S1 and S2[1:]. We still need to consider all of the characters of S1, but we're no longer interested in S2[0] since we put a copy of it in front of S1 and thus have already taken the step of converting S1 into this particular part of S2. So,

```
option3 = -1 + alignscore(S1, S2[1:])
```

Finally, we wish to return the best of these options, which is simply max(option1, option2, option3). Therefore, our alignScore(S1, S2) function looks like this:

```
def alignScore( S1, S2 ):
    ''' Returns the sequence alignment score for S1 and S2. '''
    if S1 == '': return -1 * len(S2)
    elif S2 == '': return -1 * len(S1)
    else:
        if S1[0] == S2[0]:
            option1 = 1 + alignScore(S1[1:], S2[1:])
        else: option1 = -1 + alignScore(S1[1:], S2[1:])
        option2 = -1 + alignScore(S1[1:], S2)
        option3 = -1 + alignScore(S1, S2[1:])
        return max(option1, option2, option3)
```

Not surprisingly, this function is slow, but now that you know how to memoize recursive functions, you can (and will) make this function much faster!

8.2 From Scores to Alignments

In the previous section we saw that plum could be transformed to lemon as follows:

plum -> lum	(delete the p, 1 point penalty)
	(keep the l, 1 point reward)
lum -> lem	(change the u to an e, 1 point penalty)
	(keep the m, 1 point reward)
lem -> lemo	(insert an o, 1 point penalty)
lemo -> lemon	(insert an n, 1 point penalty)

This set of operations resulted in two symbols remaining unchanged (the l and the m) for a reward of 2, one change (u to e) for a penalty of 1, and three indels (for a penalty of 3), and thus a total score of $2 - 1 - 3 = -2$.

This transformation can be viewed another way, by placing plum above lemon, with each indel represented by a – symbol, called a *gap character*. This representation, called a *sequence alignment*, is shown below:

```
plum--
-lemon
```

Reading the alignment from left to right, we see that in the first site there is a p above and a gap character below it. This corresponds to deleting the p in our transformation. In the next position, we see the l of plum above the l of lemon. This corresponds to the l being left unchanged. Next, we see the u above the e. This corresponds to changing the u to an e. Then, we see a gap character above the o, corresponding to inserting an o. Finally, we see a gap character above the n, corresponding to inserting an n.

In the last section, we also saw the following (worse) transformation:

plum -> llum	(change the p to an l, 1 point penalty)
llum -> leum	(change the second l to an e, 1 point penalty)
leum -> lem	(delete the u, 1 point penalty)
	(keep the m, 1 point reward)
lem -> lemo	(insert an o, 1 point penalty)
lemo -> lemon	(insert an n, 1 point penalty)

This transformation left only the first l unchanged and thus had a score of $1 - 5 = -4$. It corresponds to the following alignment:

```
plum--
le-mon
```

And, just for good measure, here's a truly terrible transformation:

plum -> lum	(delete the p, 1 point penalty)
lum -> um	(delete the l, 1 point penalty)
um -> m	(delete the u, 1 point penalty)
m -> "	(delete the m, 1 point penalty)

```
" -> l              (insert an l, 1 point penalty)
l -> le             (insert an e, 1 point penalty)
le -> lem           (insert an m, 1 point penalty)
lem -> lemo         (insert an o, 1 point penalty)
lemo -> lemon       (insert an n, 1 point penalty)
```

This transformation deletes four symbols and then adds five symbols for a total score of −9. The corresponding alignment is:

```
plum-----
----lemon
```

That is indeed an alignment with a bad score. If nothing else, it demonstrates that it's always possible to transform any string to any other string by first deleting all of the symbols in the first and then adding all of the symbols of the second.

The sequence alignment representation has proven very useful to biologists. It tells us which parts of one sequence correspond to which parts of another. Or, put more formally, it tells us which parts are homologous. Why is knowing this useful? As an example, imagine a group of researchers studying muscular dystrophy, a set of muscle diseases that are frequently deadly. It turns out that mice can also get muscular dystrophy, and that the mouse and human versions can be caused by mutations in the same gene, called the dystrophin gene. Therefore, researchers have used mice as a "model" to study muscular dystrophy. They induce mutations in mice that cause the disease and then they study the mutant mice. But in order to relate the findings in mice back to humans, we need an alignment for the homologous genes. Here is a small part of such an alignment:

```
Human AACTTTCCACTGACAACGAAAGTAAAGTAAAGTATTGGATTTTTTTAAAGGGAACATGTGAATGAA
Mouse AACTTTCCACTGAGAGC----------TAGAGGATTCATTTTTTTCAA--GGAACATGCGAATGAA
```

8.3 Sequence Alignment Scoring with Variable Rewards and Penalties

As we've seen, a useful algorithm for calculating similarity scores should try to approximate the mutational processes that act on sequences. This was our motivation for considering mismatches and indels in alignScore. This principle can be taken further.

In the final homework problem for Part II, you'll be comparing protein sequences from human and chicken. When we're aligning protein sequences, we need to also consider the mutational probabilities of different amino acids. Amino acids vary in their chemical properties, and this affects the probability of one amino acid mutating into another. Consider the amino acids isoleucine, valine, and arginine. Isoleucine and valine both have hydrophobic side chains, while arginine's side chain is charged and hydrophilic. Therefore, isoleucine is more likely to mutate into valine than it is into arginine. Such effects should be taken into account in our alignment scoring. In particular, we'd like to give a higher score for pairs of amino acids that are more likely to mutate into one another.

The optional section following this one explains how these different values are computed. For now, we'll assume that we're given some negative number representing the penalty for an indel, also known as a *gap penalty*, and a substitution matrix that associates rewards and penalties for matches and mismatches, respectively. For example, when aligning protein sequences, we have 20 characters, one for each amino acid, and thus the substitution matrix is 20 × 20.

Figure 8.1 shows one commonly used substitution matrix for protein sequences called the BLOSUM 62 matrix. Notice that elements on the diagonal correspond to rewards for matches and therefore they are all positive. But, they are no longer necessarily 1 because some amino acids are more likely to remain unchanged than others. Similarly, the penalties for changes vary because, again, some changes are more likely than others. For example, isoleucine gets a penalty of −3 with arginine because the two are significantly different chemically.

You may also notice that most, but not all, of the change penalties are negative. Some penalties are 0 and some are even positive. Very strange! Why do we give a reward when we'd expect a penalty? This phenomenon is explained in the next optional section.

We can represent such a substitution matrix as a Python dictionary. For example, the BLOSUM 62 matrix (among others) is available in the Biopython package (biopython.org), where it is defined like this:

```
blosum62 = {
    ('A', 'A'): 4,
    ('B', 'A'): -2,
    ('B', 'B'): 4,
    ('B', 'C'): -3,
```

	C	S	T	P	A	G	N	D	E	Q	H	R	K	M	I	L	V	F	Y	W
C	9	-1	-1	-3	0	-3	-3	-3	-4	-3	-3	-3	-3	-1	-1	-1	-1	-2	-2	-2
S	-1	4	1	-1	1	0	1	0	0	0	-1	-1	0	-1	-2	-2	-2	-2	-2	-3
T	-1	1	4	1	-1	-1	0	1	0	0	0	-1	0	-1	-2	-2	-2	-2	-2	-3
P	-3	-1	1	7	-1	-2	-1	-1	-1	-1	-2	-2	-1	-2	-3	-3	-2	-4	-3	-4
A	0	1	-1	-1	4	0	-1	-2	-1	-1	-2	-1	-1	-1	-1	-1	-2	-2	-2	-3
G	-3	0	1	-2	0	6	-2	-1	-2	-2	-2	-2	-2	-3	-4	-4	0	-3	-3	-2
N	-3	1	0	-2	-2	0	6	1	0	0	-1	0	0	-2	-3	-3	-3	-3	-2	-4
D	-3	0	1	-1	-2	-1	1	6	2	0	-1	-2	-1	-3	-3	-4	-3	-3	-3	-4
E	-4	0	0	-1	-1	-2	0	2	5	2	0	0	1	-2	-3	-3	-3	-3	-2	-3
Q	-3	0	0	-1	-1	-2	0	0	2	5	0	1	1	0	-3	-2	-2	-3	-1	-2
H	-3	-1	0	-2	-2	-2	1	1	0	0	8	0	-1	-2	-3	-3	-2	-1	2	-2
R	-3	-1	-1	-2	-1	-2	0	-2	0	1	0	5	2	-1	-3	-2	-3	-3	-2	-3
K	-3	0	0	-1	-1	-2	0	-1	1	1	-1	2	5	-1	-3	-2	-3	-3	-2	-3
M	-1	-1	-1	-2	-1	-3	-2	-3	-2	0	-2	-1	-1	5	1	2	-2	0	-1	-1
I	-1	-2	-2	-3	-1	-4	-3	-3	-3	-3	-3	-3	-3	1	4	2	1	0	-1	-3
L	-1	-2	-2	-3	-1	-4	3	-4	-3	-2	-3	-2	-2	2	2	4	3	0	-1	-2
V	-1	-2	-2	-2	0	-3	-3	-3	-2	-2	-3	-3	-2	1	3	1	4	-1	-1	-3
F	-2	-2	-2	-4	-2	-3	-3	-3	-3	-3	-1	-3	-3	0	0	0	-1	6	3	1
Y	-2	-2	-2	-3	-2	-3	-2	-3	-2	-1	2	-2	-2	-1	-1	-1	-1	3	7	2
W	-2	-3	-3	-4	-3	-2	-4	-4	-3	-2	-2	-3	-3	-1	-3	-2	-3	1	2	11

Figure 8.1. The BLOSUM 62 substitution matrix. The matrix is 20×20, with one row and one column for each of the 20 amino acids. From Henikoff S and Henikoff JG. Amino acid substitution matrices from protein blocks. *Proceedings of the National Academy of Sciences of the USA* 1992;89(22):10915–10919.

```
('B', 'D'): 4,
('B', 'E'): 1,
('B', 'F'): -3,
('B', 'G'): -1, ... }
```

Therefore, we can look up values in the dictionary like this:

```
>>> blosum62[('A', 'A')]
4
>>> blosum62(('C', 'S')]
-1
```

Now, our new `alignScore(S1, S2, gap, substitutionMatrix)` function looks like this:

```
def alignScore( S1, S2, gap, substitutionMatrix ):
    ''' Returns the sequence alignment score for S1 and S2. '''
    if S1 == '': return gap * len(S2)
    elif S2 == '': return gap * len(S1)
    else:
        option1 = substitutionMatrix[(S1[0], S2[0])] +
                    alignScore(S1[1:], S2[1:], gap, substitutionMatrix)
        option2 = gap + alignScore(S1[1:], S2, gap, substitutionMatrix)
        option3 = gap + alignScore(S1, S2[1:], gap, substitutionMatrix)
        return max(option1, option2, option3)
```

Note that for `option1` we no longer need to check whether `S1[0]` and `S2[0]` match. If they do match, then `substitutionMatrix[(S1[0], S2[0])]` will be a positive number from the diagonal of the substitution matrix, corresponding to a reward. And, if they don't match, `substitutionMatrix[(S1[0], S2[0])]` will be the appropriate penalty value (most likely zero or negative, but occasionally a small positive number for two amino acids that are very likely to substitute for each other).

8.4 Optional Section: How Substitution Matrices are Computed

The entries in protein substitution matrices are calculated based on the probability of finding particular pairs of amino acids aligned between truly homologous sequences. The values in the BLOSUM 62 matrix (shown in Figure 8.1) were based on a set of trusted alignments taken over regions where there are no gaps. Such alignments have also been called "blocks," hence the name BLOSUM, which stands for **blocks substitution matrix**.

We can illustrate how such a matrix is made with a simple example using only two amino acids. Below is a "block" consisting of an alignment between three sequences which we'll call a, b, and c. While we've only seen how to align pairs of sequences, similar algorithms can be used to align three or more sequences.

For the moment, let's just assume that we know this example alignment to be correct.

a SCCS

b CCCS

c CCCS

The first thing we want to do is calculate a matrix of "counts" telling us how many times different pairs of amino acids are aligned with each other. We can do this by considering pairs of our three sequences (a, b, and c) in turn. Lets start with a and b. Here S is paired with C once, C with C twice, and S with S once. This seems simple enough. However, we need to consider one subtlety. S is paired with C once. But where should we put this in our matrix of counts? That matrix has an S-C entry, and also a C-S entry. To choose one or the other would be arbitrary, so our solution is to put a count of 0.5 in each. When it comes to counting an amino acid paired with itself, we don't have this problem and can add a full count. Thus the counts for the first pair of sequences, a and b, are:

```
      C       S
C   2.000   0.500
S   0.500   1.000
```

When we extend this to all possible pairs of sequences (a-b, a-c, and b-c), we get the following count matrix for our example:

```
      C       S
C   7.000   1.000
S   1.000   3.000
```

Our next step is to calculate frequencies of occurrence based on these counts. We divide each count by the sum of all the values in the count matrix, which is 12 in this case. For C paired with C we get 7/12, which is the frequency of occurrence of C aligned with C in our alignment block. The whole frequency matrix for our example looks like this:

```
      C      S
C  0.583  0.083
S  0.083  0.250
```

Ultimately, our substitution matrix will be based on a comparison of these frequencies against the frequencies we would expect in alignments between non-homologous sequences. We can estimate those expected frequencies as follows. We first calculate the frequency of the C and S amino acids in the *sequences* used in our alignment blocks. Eight out of the 12 amino acids in these sequences are C, so the frequency of C is $2/3$, and 4 out of 12 are S, so the frequency of S is $1/3$. If we made random alignments between non-homologous sequences with these frequencies of C and S, then we would expect to see C aligned to C with frequency $2/3 \times 2/3$ or 4/9, and C aligned to S with frequency $2/3 \times 1/3$, and so forth. This gives us a matrix of expected frequencies for pairs of amino acids in non-homologous alignments:

```
      C      S
C  0.444  0.222
S  0.222  0.111
```

Finally, we are ready to calculate the entries of our substitution matrix. These entries are based on the ratio of the frequency of a pair of amino acids in our alignment block to the expected frequency in non-homologous alignments. This can be thought of as an "odds ratio" of homology. To get the entry for C-C, we calculate this ratio $0.583/0.444 = 1.313$. Next we take the log base 2 of that value, and multiply by 2. So, we have, $2 \times \log_2(1.313) = 0.786$. Finally, we round to the nearest integer, which is 1 in this case.[1] For the example we have been considering, this procedure gives us the following BLOSUM-like substitution matrix

```
      C       S
C  1.000  -3.000
S  -3.000  2.000
```

Substitution matrices can be tailored for different alignment problems. If we're aligning distantly related sequences we should use a different matrix than if

---

[1] By transforming with the logarithm, we make it possible to sum scores from different alignment columns, which is what our alignment algorithm requires. We round to integers because computers operate on integers much more quickly than on decimal values.

we're aligning closely related sequences. BLOSUM 62 is actually only one of a family of related matrices. Some members of this family are suitable for closely related sequences, and some for more distantly related sequences. These different BLOSUM matrices were created by giving sequences weights in the counting step. Different weighting schemes produce different matrices. BLOSUM 62 falls in the middle of the spectrum, being best for sequences which are of intermediate distance from each other. For this reason it is used as a default in applications such as NCBI BLAST.

Recall that, in addition to substitution penalties, alignment algorithms also penalize for indels or "gaps." Different substitution matrices require different gap penalties. The optimal values have typically been determined empirically. This is done by testing various combinations of substitution matrices and gap penalties on sets of test proteins where the correct alignments are known.[2] By performing alignments with these test sets, it is possible to determine which gap parameters give the best results with a particular substitution matrix.

Finally, in our example above we simply assumed that we had a trusted set of alignments and could use them to derive a substitution matrix. But how do we get those alignment blocks? In order to make a good alignment between protein sequences, we should use a substitution matrix. But this is the thing we were trying to make in the first place! One solution to this problem is to apply the approach we have described iteratively. For example, one can begin by making blocks using a simple match-mismatch scoring scheme, and follow the procedure through to the end producing a substitution matrix. Then begin again using the newly created substitution matrix to remake the blocks, make a new substitution matrix from these, and so on. After only a few iterations, the substitution matrix made from this process converges, that is, it stops changing significantly, and we can stop the procedure and keep the resulting matrix.

8.5 Optional Section: Getting the Actual Sequence Alignment

The sequence alignment score function returns a single number but doesn't show the actual optimal alignment. An alignment identifies similarities that

---

[2] Information on the structure of proteins can be used to construct very high-quality alignments. Such structural alignments represent a kind of gold standard for protein alignments. The test data sets described are typically proteins with a known structural alignment.

can be useful in identifying genes and other functionally important parts of the sequence. Our goal in this section, therefore, is to write a new function, called `align(S1, S2, gap, substitutionMatrix)`, that returns a tuple containing both the sequence alignment score and the two strings involved in an alignment. For example, using a gap penalty of 1 and a simple substitution matrix that gives a reward of 1 for each match and a penalty of 1 for each mismatch we get:

```
>>> align('SPAMD', 'ESQPAM', 1, unitMatrix)
(1, '-S-PAMD', 'ESQPAM-')
```

Once we have those three items, we'll be able to display them in a nicer way using a helper function to get output that looks like:

```
Sequence alignment score: 1
-S-PAMD
ESQPAM-
```

We'll save the `align` function as a programming exercise for you at the end of this section, but we'll develop several similar functions here to show you the big idea.

Let's start with the `change(amount, denominations)` function that we wrote in Chapter 6. Recall that this function returns the least number of coins from the `denominations` list needed to make up the given `amount`. Our objective now is to write a function called `showChange(amount, denominations)` that returns both the least number of coins *and* a list of the coins themselves.

For example:

```
>>> showChange(48, [60, 28, 24, 12, 6, 3, 1])
[2, [24, 24]]

>>> showChange(48, [25, 10, 5, 1])
[6, [25, 10, 10, 1, 1, 1]]
```

Notice that `showChange` is returning a list whose first element is the number of coins in the optimal solution and whose second element is a list containing the optimal solution itself (i.e., it contains a set of coins whose sum is the desired amount).

We'll write our showChange function by modifying the change function, which, for reference, is given again below:

```
def change( amount, denominations ):
    ''' Returns the least number of coins required to make the given
        amount using the list of provided denominations.'''
    if amount == 0: return 0
    elif denominations == []: return float('infinity')
    elif denominations[0] > amount: return change(amount, denominations[1:])
    else:
        useIt = 1 + change(amount - denominations[0], denominations)
        loseIt = change(amount, denominations[1:])
        return min(useIt, loseIt)
```

The good news is that relatively little needs to change in change! What does need to change slightly is what we return: Each time that change returns a number, showChange needs to return a list where the first element in the list is the number of coins used (which is exactly what change returns) and the second element is a list containing the actual set of coins used.

Let's start with the first base case in change:

```
if amount == 0: return 0
```

In showChange, we'll instead return [0, []] because we use 0 coins and the set of coins is empty. In the second base case, rather than returning just the value float ('infinity') we'll now return [float('infinity'), []] because, again, we use no coins in this case.

Now, consider the next place that change can return a value:

```
elif denominations[0] > amount: return change(amount, denominations[1:])
```

In this case, we are removing denominations[0] from consideration and we can simply replace change by showChange. So far, showChange looks like this:

```
def showChange( amount, denominations ):
    ''' Takes an integer amount and a list of possible
        denominations. Returns a list containing two elements. First
        the minimum number of coins required, and second a list of the actual
        coins used in that solution.'''
    if amount == 0: return [0, []]
    elif denominations == []: return [float('infinity'), []]
    elif denominations[0] > amount: return showChange(amount, denominations[1:])
```

So far, so good! But things get just a bit more interesting in the remaining general cases.

Let's begin with the use-it case in the change function.

```
useIt = 1 + change(amount - denominations[0], denominations)
```

In this case, we're exploring the option of using a coin of value denominations[0] and thus we add 1 (indicating one coin used) to the least number of coins needed to make up the remaining amount, which is given by the recursive call change (amount - denominations[0], denominations).

But we can't simply imitate this in showChange by saying:

```
useIt = 1 + showChange(amount - denominations[0], denominations)
```

Why not? Recall that showChange returns a list and we can't add a number and a list. So, instead we break this up into a few steps in showChange as follows:

```
useItRecursiveCall = showChange(amount - denominations[0], denominations)
useItNumCoins = useItRecursiveCall[0]
useItCoins = useItRecursiveCall[1]
```

We're not done yet, but let's see what we've done so far. The first line above makes a recursive call to find the best solution assuming we dispense one coin of value denominations[0]. This is a list whose first element is the number of coins used and whose second element is another list of the actual coins used.

For example, if our amount is 42 and our denominations list is [25, 10, 5, 1] then we're exploring the case that we use a coin of value 25 and thus our recursive call will be showChange(42 - 25, [25, 10, 5, 1]), which will return the list [4, [10, 5, 1, 1]] indicating that the 42-25 = 17 amount of change can be best made up with 4 coins with values 10, 5, 1, 1.

The second line of our code above assigns the variable useItNumCoins to the number of coins (4 in our example) and the third line assigns the variable useItCoins to the list containing those coins (which is [10, 5, 1, 1] in our example).

Therefore, the number of coins in the use-it case is 1 more than `useItNumCoins` and the list of coins for the use-it case is the `denominations[0]` coin added to the list `useItCoins`. In our example, we wish to add a 25 unit coin to the list `[10, 5, 1, 1]`. So, we can define a variable called `useItSolution`, which is the list `[useItNumCoins + 1, [denominations[0]] + useItCoins]`.

Next, we look at the lose-it case. In this case, we simply drop the coin with value `denominations[0]` from further consideration. That is, we ask for `showChange(amount, denominations[1:])`. This will be a list containing the number of coins and the list of coins in the lose-it solution. Unlike the use-it case, we don't need to modify this solution by adjusting the number of coins used or changing the list of coins. So, we simply do

```
loseItSolution = showChange(amount, denominations[1:])
```

Finally, we need to decide which solution is better, the `useItSolution` or the `loseItSolution`. That all depends on which solution uses fewer coins and we find that by comparing `useItSolution[0]` with `loseItSolution[0]` (if there's a tie, we can break it arbitrarily). So, our final program looks like this:

```
def showChange( amount, denominations ):
    ''' Takes an integer amount and a list of possible
        denominations. Returns a list containing two elements. First
        the minimum number of coins required, and second a list of the
        actual coins used in that solution.'''
    if amount == 0: return [0, []]
    elif denominations == []: return [float('infinity'), []]
    elif denominations[0] > amount: return showChange(amount, denominations[1:])
    else:
        useItRecursiveCall = showChange(amount - denominations[0], denominations)
        useItNumCoins = useItRecursiveCall[0]
        useItCoins = useItRecursiveCall[1]
        useItSolution = [useItNumCoins + 1, [denominations[0]] + useItCoins ]
        loseItSolution = showChange(amount, denominations[1:])
        if useItSolution[0] <= loseItSolution[0]:
            return useItSolution
        else:
            return loseItSolution
```

Next, we should memoize this function so that it runs in a reasonable amount of time, even for large inputs. We'll need to add a memo dictionary as an input to the function and we'll need to store the solutions to problems that we've solved. The keys in the memo dictionary will be tuples containing an amount and the list of available denominations. Since a key needs to be a single entity, we'll need to combine those two items into a single tuple. *Remember, dictionary keys cannot be lists and cannot contain lists.* So, the denominations input must now be a tuple too.

The values associated with these keys will be solutions returned by the function, and those solutions are lists of the form [NumCoins, CoinList] where CoinList is itself a list. It's not a problem that we're using lists at this end – the values in a dictionary are perfectly happy to be lists and contain lists – only the keys must be list-free.

So, the final memoized version of showChange, which we'll call memoizedShowChange, looks like this, where we've highlighted the code for memoization in red.

```
def memoizedShowChange(amount, denominations, memo):
    ''' Takes an integer amount and a tuple of possible denominations.
        Returns a list containing two elements. First the minimum number
        of coins required, and second list of the actual coins used in
        that solution.'''
    if amount == 0: return [0, []]
    elif denominations == (): return [float('infinity'), []]
    elif (amount, denominations) in memo: return memo[(amount, denominations)]
    elif denominations[0] > amount:
        solution = memoizedShowChange(amount, denominations[1:], memo)
        memo[(amount, denominations)] = solution
        return solution
    else:
        useItRecursiveCall = memoizedShowChange(amount - denominations[0],
denominations, memo)
        useItNumCoins = useItRecursiveCall[0]
        useItCoins = useItRecursiveCall[1]
        useItSolution = [useItNumCoins + 1, [denominations[0]] + useItCoins ]
        loseItSolution = memoizedShowChange(amount, denominations[1:], memo)
```

```
        if useItSolution[0] <= loseItSolution[0]:
            solution = useItSolution
        else:
            solution = loseItSolution
        memo[(amount, denominations)] = solution
        return solution
```

8.6 Identifying Orthologs

The sequence alignment score function represents a good measure of sequence similarity. Next we need a way to use this measure to recognize orthologs.

The first principle we'll use is that true orthologs are more similar to each other than non-orthologs. True orthologs have actually descended from a common ancestor, and the differences between them can be explained by mutational events that occurred during their evolutionary history. For mammal and bird proteins, true orthologs are typically separated by a moderate number of mutations. When we pass a pair of proteins like this to `alignScore`, we get a high score back, reflecting the fact that `alignScore` can convert one protein to the other using only a small number of mismatches and indels. In contrast, non-orthologs are not actually descended from a common ancestor. It is possible to convert one string into the other; however, it requires relatively many mismatches and indels, and as a result `alignScore` returns a low score.

This still leaves an important question: How high a score is high enough for us to conclude that two proteins are orthologs? Trying to establish a cutoff could be tricky. We'll take an alternate approach called the method of *best reciprocal hits*.

The best reciprocal hits method works like this. Imagine that we have a collection of human genes and a separate collection of chicken genes. A small example with five human genes and six chicken genes is shown in Figure 8.2. Consider a specific human gene that we'll call h1. We compute the sequence alignment score between h1 and every chicken gene in our data set. Among those chicken genes, imagine that gene c3 has the highest sequence alignment score. So, it seems that h1 and c3 are similar. But, we haven't yet compared c3 to other human genes. So, next we do that. Imagine that among all the human genes, it turns out that h1 has the

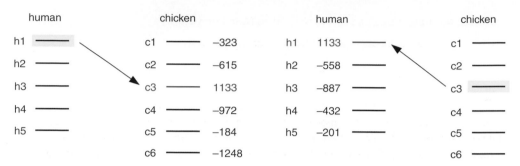

Figure 8.2. An example of the best reciprocal hits method. In the left panel we compare the human gene h1 with the set of chicken genes. Gene c3 has the highest sequence alignment score. On the right panel we search the set of human genes with chicken gene c3. The human gene h1 has the highest sequence alignment score, making h1 and c3 best reciprocal hits.

highest sequence alignment score with c3. In this case h1 and c3 are said to be best reciprocal hits. The "reciprocal" here means that it's mutual – h1 finds that c3 is the most similar chicken gene *and* c3 finds that h1 is the most similar human gene. Such a relationship is often taken as evidence that genes are orthologs.

8.7 Comparing Chromosomes

So far, we've described a method to determine which genes are orthologs. But how do we determine whether two chromosomes are likely to be homologous? Our first inclination might be to use the sequence alignment score function to compare pairs of chromosomes. Although this function will be much faster once it's memoized, the human X chromosome is more than 150 million base pairs long and even the memoized function will take considerable time to analyze sequences of this length. But an even bigger problem is that while gene evolution can be approximated by changes, insertions, and deletions, chromosome evolution involves additional mutational processes that are not adequately captured by the sequence alignment score function. For example, there are several types of mutations that move large pieces of DNA from one chromosome to another. Figure 8.3 shows an example of one such kind of mutation called a *reciprocal translocation.*

Since the time of their last common ancestor about 300 million years ago, birds and mammals have been evolving independently. When a mutation like a reciprocal translocation occurs, it rearranges the chromosomes in one species but not the other. Over such a long period of time, many rearrangements are likely to have occurred. Thus, when we compare the chromosomes in mammals and in birds,

Figure 8.3. A reciprocal translocation event. We begin with two different chromosomes in the same species, each with its own set of genes. When a reciprocal translocation occurs, the two chromosomes break and swap ends. As a result, the combination of genes present on a single chromosome changes.

it's very unlikely that we'll find whole chromosomes that are entirely homologous with each other. Instead, we're only likely to find pieces of chromosomes that are homologous between the two species.

Figure 8.4 shows two possible scenarios for the evolution of the mammalian and the bird sex chromosomes. The first scenario represents a case where the mammalian X and the bird Z chromosomes are partially homologous. The common ancestor is shown with four colored chromosomes (of course there were more, but we only show four for simplicity). In the lineage leading to mammals, several translocations occurred, causing the pieces of the ancestral chromosomes to be shuffled. In the lineage leading to birds, an entirely different translocation occurred. In this scenario, when we try to compare the different chromosomes, we find that part of the mammalian X chromosome is homologous to the bird Z chromosome. But another part of the mammalian X chromosome is homologous to a different bird chromosome.

The second scenario shows what we might expect if the X and the Z chromosomes are not homologous at all. As in the top case, chromosomal rearrangements have occurred since the time of the common ancestor. But when we compare the X and Z chromosomes we find that they don't have any regions that are similar to each other.

Chromosomal rearrangements are the reason that our strategy involves comparing genes rather than whole chromosomes. For each gene on the human X chromosome we seek the orthologous gene in chicken and determine which chicken chromosome contains that gene. If none of the genes on the human X chromosome have orthologs on the chicken Z chromosome, then we can conclude that these two chromosomes are not homologous (scenario 2). If instead there are some genes with homology, then we'll conclude that they're partly homologous (scenario 1).

Scenario 1: X and Z partially homologous

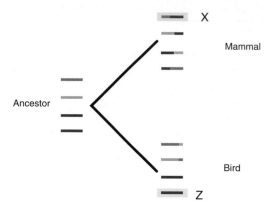

Scenario 2: X and Z not homologous

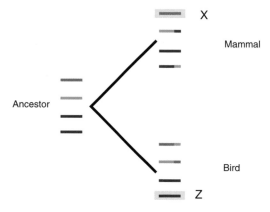

Figure 8.4. Two simple scenarios for sex chromosome evolution. In the top scenario, the mammalian X and the bird Z chromosomes are partially homologous. In the bottom scenario, they are not homologous.

Putting it All Together

The sequence alignment technique that we discussed in this chapter is a fundamental method in computational biology. It is used in hundreds if not thousands of different pieces of software for tasks such as finding sequence similarity in a database or constructing large multiple alignments. Most of these programs use algorithms related to the one we have described here.

This chapter concludes Part II, where we have focused on the concept of recursion: Functions calling themselves for help. We've seen how the use-it-or-lose-it paradigm can help us write recursive functions, and how some problems that are hard to solve in other ways are easy to solve using this technique. We've also learned about the dictionary data type, and seen how it can be used with the memoization technique to speed up recursive functions.

In the final programming problem for Part II, you'll write a program that compares the genes on several human and chicken chromosomes, and uses the results to determine the relationship between the human and chicken sex chromosomes!

FOR FURTHER READING

Livernois AM, Graves JA and Waters PD. The origin and evolution of vertebrate sex chromosomes and dosage compensation. *Heredity* 2012;108(1):50–58.

Wilson MA and Makova KD. Genomic analyses of sex chromosome evolution. *Annual Review of Genomics and Human Genetics* 2009;10:333–354.

PART III

Phylogenetic Reconstruction and the Origin of Modern Humans

· ·

Sixty thousand years ago, southwestern France was in the midst of an ice age. The climate was much cooler than it is today, and the region had different animals. These included many large, cold-adapted animals such as woolly mammoths, Irish elk, and cave bears.

There were also people living in the region. Artifacts from this time are abundant, and include spear points and scrapers of the kind that would be used in killing and butchering big game. Human remains reveal people with a short, stocky build, and distinctive facial features including a prominent nose and swept back cheeks. These people were Neanderthals, and by this time they had been occupying Europe for many millennia (Figure 9.1).

Since that time, Europe came to be occupied by other humans, modern humans like ourselves. A key issue in human evolution is the connection between later humans and Neanderthals. Are modern humans descendants of Neanderthals? Or did they come from elsewhere and replace them? Computational approaches have made key contributions toward answering this fascinating question.

The story of Neanderthals and modern humans takes place in the last several hundred thousand years. But let's step back for a moment and put this in the larger perspective of human evolution. Species more closely related to modern humans than to our closest living relatives (chimpanzees) are called *hominins*. The hominin fossil record is quite extensive, and from it paleontologists have learned many things about human origins. One conclusion is that hominin evolution likely began in Africa. From the time of the divergence with chimpanzees 5–8 million years ago until about 2 million years ago, hominin fossils are all found inside Africa. It is only about 2 million years ago that we begin to find evidence of hominins outside Africa.

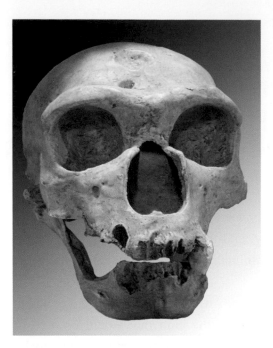

Figure 9.1. The old man of La Chapelle. This Neanderthal fossil was found in a cave in southwestern France and dates to between 40,000 and 60,000 years ago. © Luna04/Wikimedia Commons/CC-BY-SA-3.0/GFDL/http://en.wikipedia.org/wiki/File:Homo_sapiens_neanderthalensis.jpg).

The first undisputed world traveler was a species called *Homo erectus*, which represents something of a milestone in human evolution. In terms of stature and gait, *Homo erectus* was similar in many ways to modern humans. Their brains were smaller than ours, but they were sophisticated in their use of stone tools and in other ways. They were able to adapt to a wide variety of environments, ranging from warm weather and tropical environments in Africa, to much cooler and/or drier sites in Asia and Europe.

Homo erectus is widely believed to be the ancestor of later hominins such as Neanderthals and modern humans. There has been disagreement, however, about how exactly *Homo erectus* led to those later species. One view, called the *multi-regional model*, suggests that local populations of *Homo erectus* in different parts of the world independently evolved into modern human populations. *Homo erectus* in Asia would have led to modern Asian people, and *Homo erectus* in Europe would have evolved into modern Europeans. In this view, Neanderthals were an intermediate stage in that evolution, and thus were ancestors to modern Europeans.

An alternative idea is that modern humans evolved in Africa, and spread throughout the world from there. This is called the *out-of-Africa model* and in this scenario indigenous local populations of *Homo erectus* or its descendants were replaced by modern humans spreading from Africa. If this scenario is correct, then Neanderthals did not evolve into modern Europeans, but rather were replaced by them.

These questions are fundamentally about the relationships between biological species. The discipline of reconstructing such relationships is called *phylogenetics*. In this part of the book, we will study general computational approaches to reconstructing the evolutionary history, or *phylogeny*, of a set of species. In the final homework problem we'll apply these methods to determining the relationship between Neanderthals and modern humans.

Representing and Working with Trees

9

Computing content:
representing phylogenetic trees

Biologists are frequently interested in the relationships between species. To represent such relationships they use a diagram called a *phylogenetic tree*. The phylogenetic tree in Figure 9.2 consists of a set of nodes connected by branches. In this example, we have marked the nodes with dots. The ones on the right are called *leaf nodes* or simply *leaves*. They represent the species whose relationships we want to understand. In this example, the leaves are five currently living primate species, including great apes and humans. There are also internal nodes, which represent hypothesized ancestral species. For example, the node marked in red represents the most recent common ancestor of chimpanzees and humans, an animal that lived between 5 and 8 million years ago.

The tree provides information about the evolutionary relationships between species. We can use the internal nodes to define groups of closely related species called *clades*. All the species descended from a particular internal node form a clade, and are more closely related to each other than they are to other species. Thus the red dot in Figure 9.2 defines a clade that we might call the human-chimpanzee clade. This clade includes three living species: human, the common chimpanzee, and the pygmy chimpanzee. The fact that they are in a clade together tells us, for example, that chimpanzees are more closely related to humans than they are to gorillas.

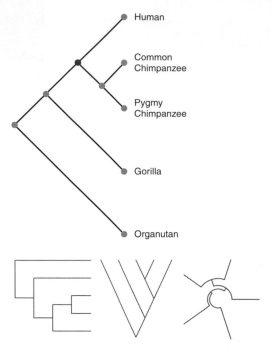

Figure 9.2. (Top) A phylogenetic tree of primates. The node marked in red represents the most recent common ancestor of chimpanzees and humans. (Bottom) Several alternate representations of phylogenetic trees.

The tree in Figure 9.2 has living species as its the leaves. However, phylogenetic trees can also include fossil species at the leaves. In the final homework problem for this part you will be including the fossil species Neanderthal in the tree that your program constructs.

We've drawn our tree with angled branches and with the leaves on the right. However, trees can be drawn in a variety of ways, some of which are illustrated at the bottom of the figure.

9.1 Representing Trees

In order to construct, draw, and analyze phylogenetic trees, we'll need a convenient representation of trees that can be manipulated in our computer programs. At first glance, it seems that representing a tree might be messy and complicated because trees branch in all kinds of ways. Surprisingly, though, there's a simple and elegant way to represent trees as lists or tuples.

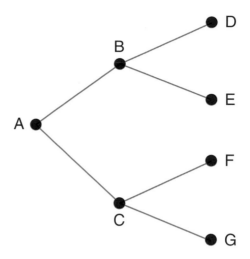

Figure 9.3. A small phylogenetic tree.

To get started, consider the small tree shown in Figure 9.3.

Recall that the nodes labeled D, E, F, and G at the far right of the tree are called *leaves*. (Biologists also sometimes call them *tips*.) And all of the other nodes are called *internal nodes*. The internal node at the far left, which we've labeled A, is called the *root* of the tree. Moreover, each internal node is the root of its own smaller tree. For example, the internal node B defines a smaller tree that contains nodes D and E. In the terminology we introduced above, this smaller tree represents a clade. Another term for it, frequently used by computer scientists, is *subtree*.

We use family tree metaphors to describe the relationships in a tree. Each internal node has two *children* and each child, other than the root, has a unique *parent*. (Throughout the remainder of this chapter, we assume that a node has either two children or no children. In the former case the node is an internal node and in the latter case it's a leaf.) Continuing with the family tree analogy, the *descendants* of an internal node are all of the nodes in its subtree. For example, the descendants of A are B, C, D, E, F, and G while the descendants of C are just F and G.

A tree is a recursive structure: It's made of a root and two subtrees. So, the tree in Figure 9.3 can be represented by a tuple (or a list) of the form (`'A'`, `SubtreeA1`, `SubtreeA2`). What are `SubtreeA1` and `SubtreeA2`? They are trees themselves! So, `SubtreeA1`, for example, can be represented as a tuple (`'B'`, `SubtreeB1`, `SubtreeB2`) where `SubtreeB1` and `SubtreeB2` are, again, trees. But is `SubtreeB1` really a tree? Its root is D but it seems to have no subtrees. In fact, D *does* have subtrees, but they're empty! So, SubtreeB1 can be represented as ('D', (), ()). Similarly, SubtreeB2 is represented as ('E', (), ()). Finally, the tree in Figure 9.3 is represented as the tuple below that we've named smallTree. We've split the tuple over multiple lines and used indentation to help you see the structure. (Python is fine with extra lines and indentation, so we recommend using this format when you define your own trees.)

```
smallTree = ('A',
              ('B',
                ('D', (), ()),
                ('E', (), ())
              ),
              ('C',
                ('F', (), ()),
                ('G', (), ())
              )
            )
```

Notice that the length of smallTree is 3 (i.e., if we type len(SmallTree), Python returns 3). It seems to have more than three things in it, but in fact it's a tuple with three "things," some of which happen to be tuples themselves. In general, according to our method of representation, a tree comprises three things: A root and two subtrees.

We've seen that a tree can be represented by a tuple of the form (Root, Subtree1, Subtree2) or an empty tuple (). You might find it weird that we use the empty tuple at all. Why do that? We could simply decide that leaves are represented as a string and be done with it. That would indeed be another valid way of representing trees. However, by using the empty tuple we can recognize leaves particularly easily: A leaf is a tree in which both of its subtrees are empty tuples.

9.2 Computing with Trees

In this section, we'll write some short Python functions that "do things" with trees. These functions are useful in their own right and will also help us develop programming skills that we'll need later.

To get started, imagine that we're given a tuple representing a tree and wish to count and return the total number of nodes in that tree. Our function, which we'll call `nodeCount( Tree )`, takes a tree as input and returns a number as output. For example, for the `smallTree` example above, this is what the function should do:

```
>>> nodeCount(smallTree)
7
```

Since a tree is a recursive structure, our `nodeCount` function will, naturally, be recursive too. The number of nodes in a tree is just the number of nodes in each of the two subtrees plus one to account for the root. So, to count the nodes in a tree, we'll need to be able to count the number of nodes in a subtree. But counting the number of nodes in a subtree is precisely what the `nodeCount` function does, so we can apply it recursively to each of the two subtrees. We'll need a base case too, but for now we can capture this idea in this not-quite-complete function:

```
def nodeCount( Tree ):
    ''' Compute the total number of nodes in a given Tree.'''
    Subtree1 = Tree[1]
    Subtree2 = Tree[2]
    return 1 + nodeCount(Subtree1) + nodeCount(Subtree2)
```

Remember that the input `Tree` is of the form `(Root, Subtree1, Subtree2)`, so `Tree[1]` corresponds to the first subtree and `Tree[2]` corresponds to the other subtree.

Now for the base case. We ask ourselves, "what kind of input tree would allow us to return an answer right away?" If the tree is a leaf, that is, both of its children are empty tuples, then there's just one node in that tree and we should return 1. There is no need to make a recursive call. Keep in mind that in the trees that we're considering here, a node cannot have just one child, so either both of its children are empty tuples or neither are empty tuples. So, it suffices to test whether just one of the two children is empty and, if so, return 1. Our completed program looks like this:

```
def nodeCount( Tree ):
    ''' Compute the total number of nodes in a given Tree.'''
    Subtree1 = Tree[1]
    Subtree2 = Tree[2]
```

```
    if Subtree1 == (): return 1 # It's a leaf!
    else:
        return 1 + nodeCount(Subtree1) + nodeCount(Subtree2)
```

Let's consider what happens when we run this program on our `smallTree` example. In this case, `Subtree1` is:

```
('B',
    ('D', (), ()),
    ('E', (), ())
)
```

and `Subtree2` is:

```
('C',
    ('F', (), ()),
    ('G', (), ())
)
```

Since `Subtree1` is not an empty tuple, the base case doesn't apply, and our function returns

```
1 + nodeCount(Subtree1) + nodeCount(Subtree2)
```

Python now first computes `nodeCount(Subtree1)` and, when that's done, continues on to return `nodeCount(Subtree2)`. Let's just pursue the computation of `nodeCount(Subtree1)`. In this case, `nodeCount` gets called again with the input tree:

```
('B',
    ('D', (), ()),
    ('E', (), ())
)
```

Again, the base case doesn't apply, so this function returns

```
1 + nodeCount( ('D', (), ()) ) + nodeCount( ('E', (), ()) )
```

When `nodeCount( ('D', (), ()) )` is called, the base case applies because the first subtree is (). So, this returns 1. Similarly, `nodeCount( ('E', (), ()) )` returns 1. Thus, the expression...

```
1 + nodeCount ( ('D', (), ()) ) + nodeCount ( ('E', (), ()) )
```

... now returns $1 + 1 + 1 = 3$. Similarly, the `nodeCount` of the tree...

```
('C',
    ('F', (), ()),
    ('G', (), ())
)
```

... also returns 3. And finally, the original recursive call returns $1 + 3 + 3 = 7$, which is the answer we were looking for – the number of nodes in the original tree.

Let's try a few more examples. The *height* of a tree is the number of branches along the longest path from the root to a leaf. For example, our `smallTree` has height 2 because the longest path takes two steps to get to a leaf. In this particular tree, every path from the root to a leaf has length 2. In the tree in Figure 9.4, however, not all paths from the root to the leaves have the same length but the longest such path has length 3, so the height is 3.

The height of a tree depends on the heights of its two subtrees. In Figure 9.4, the top subtree of the root has height 0 while the bottom subtree has height 2. So, the height of the entire tree is $1 + 2$: The 2 accounts for the height of the taller of the two subtrees, and the 1 accounts for the branch from the root to that subtree. In general, unless the tree is a leaf (which will again be our base case), the height of the tree is one plus the larger of the heights of its two subtrees. Note that the larger of two or more values can be found using Python's built-in `max` function.

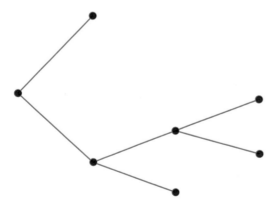

Figure 9.4. A tree with height 3.

For the base case, we need to check whether the tree is a leaf, that is, whether one of its children is empty. If so, the height is 0 since there are no branches emanating from a leaf. So, our `height` function now looks like this:

```
def height ( Tree ) :
    ''' Compute the height of a given Tree.'''
    Subtree1 = Tree [1]
    Subtree2 = Tree [2]
    if Subtree1 == () : return 0  # It's a leaf!
    else:
        return 1 + max (height (Subtree1), height (Subtree2))
```

Let's do one more example. Imagine that we want to collect the list of all leaves in a tree. That is, we want the list of all currently living species represented in that tree. This `leafList` function will return a list rather than a number, so both the base case and the general case must return lists.

If the tree is a leaf, we simply return the *list* containing that single leaf. Otherwise, we want to find the list of leaves for each of the two subtrees and return a new list that is the concatenation of those two lists. This, you may recall, can be found by simply adding the two lists. So, our function looks like this:

```
def leafList ( Tree ) :
    ''' Returns the list of leaves in a given Tree.'''
    Root = Tree [0]
    Subtree1 = Tree [1]
    Subtree2 = Tree [2]
    if Subtree1 == () :  return [Root] # Return the list with one item
    else:
        return leafList (Subtree1) + leafList (Subtree2)
```

Putting it All Together

Robert Frost's poem "The Sound of Trees" begins with the line: "I wonder about trees." At this point, you might be wondering why we've spent time counting their leaves, among other tasks. As we noted, trees are useful in representing evolutionary histories and that's where we're headed. In the next chapter, we'll learn how to write recursive programs to draw trees and, from there, we'll move on to inferring evolutionary trees from biological data. Ultimately, we'll use these methods to explore the evolutionary histories of modern humans and Neanderthals.

10 Drawing Trees

> **Computing content:**
> drawing phylogenetic trees

One of the things that distinguishes anatomically modern humans from other human groups in the fossil record is the diversity and quality of the artifacts they produced. Figure 10.1 shows three bone needles that were discovered in Liaoning province in northeastern China. These date to between 20,000 and 30,000 years ago, and were almost certainly produced by anatomically modern humans. Liaoning province is cold today, and was even colder tens of thousands of years ago. Needles like these were used to produce individually tailored sewn clothing, which is very warm.

The Liaoning site shows that anatomically modern humans were living at the far eastern edge of Asia by 20,000 to 30,000 years ago. A very important question is how they got there. Amazingly enough, this question can be addressed through the analysis of sequences taken from living humans.

How is this possible? The solution involves building a phylogenetic tree from the sequences, and then making inferences based on this tree. We'll illustrate this with an example.

Remember that one important theory of modern human origins is the out-of-Africa model, which posits that humans originated in Africa and then spread rapidly to the rest of the world. If this model is correct, it should be reflected in the relationships between modern human groups. Figure 10.2 shows a hypothetical

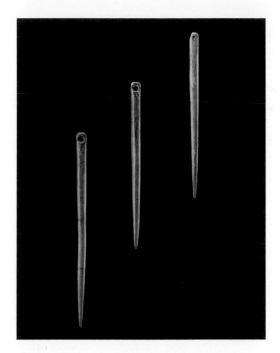

Figure 10.1. Three needles made of bone from Liaoning province, China. They date to between 23,000 and 30,000 years before present. Image source: Human Origins Program, Smithsonian Institution (http://humanorigins.si.edu/evidence/behavior/bone-and-ivory-needles), used with permission.

scenario where modern humans originated in Africa and migrated outward from there (in reality, even if out-of-Africa is correct, the migration probably didn't look exactly like this). They began somewhere around the gray dot. At this location, some humans stayed put, while others moved northeast. Some of those that moved settled a little bit further on (red dot), while others continued to the northeast. Among those that continued, some settled in Central Asia (blue dot), and some pushed further east still (green dot). This pattern of migration implies that among the groups shown, green and blue should be the most closely related, with red being next closest to them and gray the most distantly related from the others. These relationships are shown in the phylogenetic tree on the bottom, which gives an indication of what we would expect if the out-of-Africa model is correct. The important point is that the earliest diverging groups in the tree are African. This is what would result if there was a rapid migration from an African homeland. Alternative models, such as the multiregional model, would not produce this pattern of relationships.

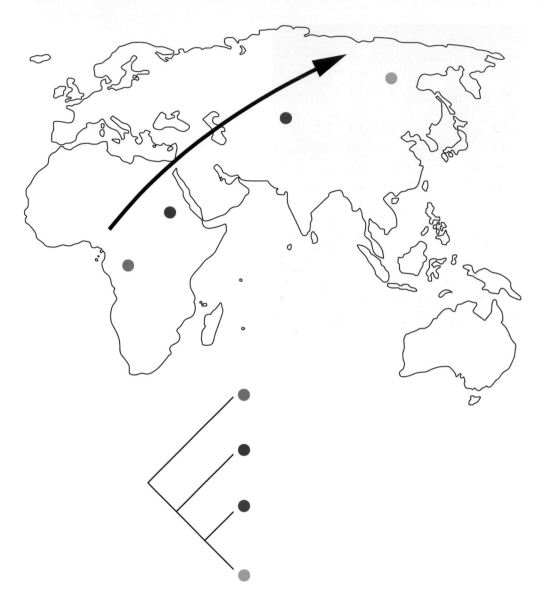

Figure 10.2. Phylogeny and the out-of-Africa model: a hypothetical example.

In the final homework problem, you'll study the geographic origin of modern humans by making your own phylogenetic trees. The algorithm for this (discussed in Chapter 11) produces trees in the tuple form we have used already. But a tuple is not a very useful thing to look at. To make our output usable, we need a good way to visualize it. That's the subject of this chapter.

Figure 10.3. Screenshot after performing `turtle.forward(100)`.

Python offers a number of different ways to draw in 2D and 3D. We'll draw trees using Python's `turtle` package, which provides a virtual turtle that draws lines as it moves under our control. While we'll ultimately control the turtle through a program, it's worth playing with our new turtle friend for a few minutes first. To that end, load the turtle package into the Python interpreter like this:

```
>>> import turtle
```

Now, you can issue commands to tell the turtle to move forwards, backwards, turn left, and turn right. For example, after typing…

```
>>> turtle.forward(100)
```

… you'll see something like what's shown in Figure 10.3 on your screen.

Next, we can turn the turtle right 90 degrees this way:

```
>>> turtle.right(90)
```

Before reading on, try drawing some nice shapes using `forward`, `right`, `left`, and `backward`. There are more turtle commands available too and you can learn about these by searching online for "python turtle graphics."

It's more interesting to use the turtle under a program's control. For example, here's a function that draws a square whose side length is specified by user input:

```python
import turtle # include this at the top of your file

def square( sideLength ):
    ''' Draws a square of given side length. '''
    for num in range(4):
        turtle.forward(sideLength)
        turtle.right(90)
```

10.1 Drawing Fractal Trees

Before drawing phylogenetic trees, we'll develop our drawing skills in another related context. Take a look at the tree in Figure 10.4(a). This tree is said to be *fractal* because its smaller parts are similar to the whole. The tree is made up of a trunk and two smaller copies of itself, one shown in red and the other in green in Figure 10.4(b). This tree is practically screaming "recursion!"

Our `fractalTree` function will take a `trunkLength` number as input. It will draw a trunk of that length, turn left 45 degrees, and recursively draw a tree with a trunk length that is one half of the original trunk length (as shown in the red tree in Figure 10.4(b)). Then, it will turn right and draw another identical recursive tree.

Of course, we need some way to tell the recursion to stop eventually, so our base case will tell the function to simply return when the trunk length becomes less than 10. (Later, you might want to modify some of the values that we've chosen here so that the angle turned is something different from 45 degrees, the recursive trunk length is not necessarily half of the original trunk length, and the base case uses some different condition to stop the recursion.)

Our recursive function looks like this, where we've indicated the two recursive calls in red and green, corresponding to the red and green subtrees shown in Figure 10.4(b):

```
import turtle

def fractalTree(trunkLength):
    ''' Recursively draw a tree given a starting trunk length. '''
    if trunkLength < 10:                    # Base case
        return
    else:
        turtle.forward(trunkLength)         # Draw the trunk
```

(a)

(b)

Figure 10.4. (a) A fractal tree. (b) The tree is made of a "trunk" (in black) and two smaller trees, one shown in red and the other in green.

```
turtle.left(45)                  # Turn left
drawTree(trunkLength/2.0)        # Draw first smaller tree
turtle.right(90)                 # Turn right
drawTree(trunkLength/2.0)        # Draw second smaller tree
turtle.left(45)                  # Turn to face forward
turtle.backward(trunkLength)     # Backup to where we started
```

The last two lines of this function deserve some closer attention. Why did we turn left 45 degrees and then go backwards at the end? Notice that this effectively returns the turtle to the position and orientation where it started. But why bother? Why not just leave the turtle in some other position and orientation?

The answer can perhaps be best captured by the quip "if parents expect good behavior from their children, they must model it themselves." In this case, the "parent" is the original call to the `fractalTree` function and the two "children" are the two recursive calls to `fractalTree`. Notice that after the first call to `fractalTree` – which we've highlighted in red in our code because it drew the red tree in Figure 10.4(b) – we want to turn the turtle right 90 degrees to orient it to draw the second tree, the green one in Figure 10.4(b). But this requires that the turtle be where we left it immediately before that first red recursive call. If the red `fractalTree` recursive call left the turtle in some strange place, we'd have to do more work to locate it and move it to the right spot before invoking the green `fractalTree` recursive call. In other words, we want each recursive call, that is, each "child," to return the turtle to where it was just before the recursive call. Remember that our children behave like we do since they are recursive copies of us (at least in the domain of programming!). Therefore, we must return the turtle to the location and orientation in which it started so that our children will do the same.

Finally, we note that Python's turtle always starts pointing "east," so we first issued the command `turtle.left(90)` before invoking the `drawTree` function so that the tree would be oriented upwards. However, if you don't mind "fallen" trees, you can skip that!

10.2 Drawing Phylogenetic Trees

In the next chapter, you'll learn an algorithm for constructing phylogenetic trees and you'll implement it in Python. That algorithm will construct trees using the tuple representation that we've used throughout this chapter. Then, you'll want to

draw those trees. In this section, we'll get you started writing the `drawPhyloTree`
function, which takes a tuple representation of a tree as input and draws that tree
using the turtle and recursion.

Let's revisit the small tree example from Chapter 9.

```
smallTree = ('A',
                ('B',
                    ('D', (), ()),
                    ('E', (), ())
                ),
                ('C',
                    ('F', (), ()),
                    ('G', (), ())
                )
            )
```

When we run. . .

```
>>> drawPhyloTree(smallTree)
```

. . . we might imagine that the turtle would draw a picture like the one shown in
Figure 10.5.

What happened here? The turtle starts pointing "east." We turned the turtle left
45 degrees and then called the `drawPhyloTree` function to draw the first subtree:
`('B', ('D', (), ()), ('E', (), ())`. When that was done, we turned the turtle
right 90 degrees and called the `drawPhyloTree` function again, this time to draw
the other subtree: `('C', ('F', (), ()), ('G', () ())`.

What was the base case that caused the drawing to stop? When we recursed on a
tree with no children (a fact that we recognized by seeing that `Tree[1]` was `()`,
which implied that `Tree[2]` was `()` as well), we printed the root of that tree. For
example, when `drawPhyloTree` finally was asked to draw the tree `('D', (), ())`,
it noticed that the first subtree was `()` and thus printed 'D' on the screen and
returned. We can draw a string on the screen using: `turtle.write(X)` where `X` is
the string that we want to print.

You're ready to write the `drawPhyloTree` function yourself and we're about to
pause here to let you do that. Before you get started, we'll point out just two things.
First, you'll notice that only the leaves of the tree are being printed. In Figure 10.5,

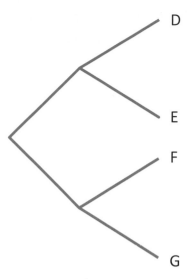

Figure 10.5. One possible way to draw the smallTree.

"D," "E," "F," and "G" were printed because they "tripped" the base case that caused them to print. But, the names of the internal nodes "A," "B," and "C" were not printed. This is actually desirable, because we generally don't have names for those interior nodes – they are hypothesized ancestral species. For that reason, we will use a "dummy" name such as 'ancestor' for all interior nodes in our trees. For example, our small tree will be represented like this:

```
smallTree = ('ancestor',
            ('ancestor',
                ('D', (), ()),
                ('E', (), ())
            ),
            ('ancestor',
                ('F', (), ()),
                ('G', (), ())
            )
        )
```

Happy drawing!

Putting it All Together

When you drew your first tree as a child, you couldn't possibly have imagined that you'd someday write a recursive function to draw a tree! But indeed, the phylogenetic trees that you see in biology books, journals, and elsewhere were drawn by recursive programs.

One of the strange things about drawing trees recursively is that, unlike other recursive functions that we've seen, these functions don't return anything. The recursion is being used to control the drawing, but there's no numerical or other result that the drawing function returns. The same is true in many other uses of recursion, among them the use of recursive functions to control a robot's exploration of an environment.

You've now learned how to represent and draw phylogenetic trees. In the final chapter of this part, you'll learn how to reconstruct phylogenetic relationships from sequence data, and apply this technique to the problem of modern human origins.

The UPGMA Algorithm

> **Computing content:**
> reconstructing phylogeny with
> the UPGMA algorithm

Our final task is to develop an algorithm to reconstruct phylogenetic relationships based on sequence data. In the final homework problem, the source sequences are mitochondrial DNA from a number of modern human individuals, as well as from several fossils, including a Neanderthal. Let us begin by saying something about these sequences and how they are used to create input for our algorithm.

The cells of eukaryotes, such as humans, contain two types of DNA. The largest type is the nuclear DNA which is found in sets of chromosomes that are inherited sexually, with one copy of each chromosome coming from either parent. A second type of DNA can be found in the mitochondria, organelles specializing in energy metabolism. Mitochondria contain their own circular DNA molecule. As it turns out, mitochondria are inherited maternally – individual humans get their mitochondria from their mother's egg rather than their father's sperm. Thus, mitochondrial DNA is passed along the maternal line only.

Mitochondrial DNA has frequently been used in studies of human evolution. One advantage is the fact that it's inherited from a single parent, and thus is not subject to recombination. Another advantage is the fact that mutations arise comparatively quickly in mammalian mitochondrial DNA. If we are comparing closely related samples, such as human individuals, a higher rate of mutation is good because it produces more differences with which to distinguish the samples.

We'll be working with mitochondrial data from a number of contemporary humans and two fossils. The first fossil is from a Neanderthal, found in Croatia, which dates to about 38,000 years ago. The second is a fossil from an anatomically modern human found in Russia, which is from about 30,000 years ago. The recovery and sequencing of DNA from fossils has been one of the most remarkable scientific accomplishments of the last several years. Having sequences from the Neanderthal allows us to directly address the relationship between this species and our own.

11.1 The Algorithm

The algorithm that we'll learn does not work on the sequences directly. Instead, it uses a distilled form of the data called a *distance matrix*. To create a distance matrix we calculate a numerical distance between each pair of sequences. For example, consider the aligned sequences below from three hypothetical species. We've colorized the bases in these sequences (A red, T blue, G black, C green) to make it easier to see where the species have differences.

A CATCAACCAGTGACCAGTATAGGACGCC

B CAACACTCAGTGACAAGTCTAGCACGCC

C AATCGCCCGGCGTCAGGCATAGCAAGCG

With a multiple alignment of this sort we can determine a distance between each pair of species. These distances are similar in spirit to the sequence alignment scores we calculated in Part II, though they typically ignore sites with gaps (in our example there happen to be no gaps).

There are various ways to calculate the distance between a pair of sequences. In this example we'll use the *Hamming distance*, which is simply the number of sites (that is, positions) where a pair of sequences differ. Consider species A and species B in the above example. We can loop over all the positions in the alignment, comparing the base in A and the base in B at each position. We find that A and B differ at a total of six sites, and thus the Hamming distance between these sequences is 6. Below is the full matrix of Hamming distances for species A, B, and C. Notice that that the distance between a sequence and itself is 0. Moreover, once we've computed the distance from A to B, we know that the distance from B to A will be the same. For this reason, we often only write the entries below the diagonal.

	A	B	C
A	0		
B	6	0	
C	12	12	0

In real applications (including the homework) distance matrices are usually made with somewhat more complicated methods that make allowances for the fact that multiple mutations might arise at the same site.

Our algorithm for phylogenetic reconstruction is called the Unweighted Pair Group Method with Arithmetic means (UPGMA). The UPGMA approach assumes that the sequences being used have evolved according to a *molecular clock*. That is, it assumes they evolved from a common ancestor, with substitutions (i.e., changes) accumulating in a regular, clock-like manner.

What does this mean exactly? Imagine that we have an ancestral species that at some point diverges into two child species. If substitutions subsequently accumulate at the same rate in those two child species, then we have clock-like evolution. But there are many things that determine the substitution rate, and it's not guaranteed to be the same in two different species. For example, the mutation rate might change, or natural selection might start acting differently in one species or the other. For this reason, the molecular clock assumption is not always justified. In some sequences, such as the human sequences we'll be looking at in the homework, it is reasonable.[1] But in other cases one needs to use a different method that doesn't rely on the molecular clock assumption.

If there has been a molecular clock governing the rate of sequence evolution, then sequence change will be proportional to time, and can be treated as a sort of measure of time. As we'll discuss below, if there is outside information available about the date of one or more of the divergences in the tree, then we can calibrate sequence changes into actual time.

The UPGMA algorithm begins with a set of leaf nodes (species) provided as input. It finds the two species that are most closely related – that is, that have the smallest distance between them – and merges them into a new node. This new node defines a clade (in biology parlance) or a subtree (in computer science speak). The

[1] This is true of mitochondrial sequences excluding the D-loop region (Ingman M et al. *Nature* 408. 2000). We exclude the D-loop from the sequences we use in the homework problem.

algorithm replaces the two merged nodes with this new single node, and does some arithmetic (described below) to compute the distance between the new node and all of the remaining nodes. It then repeats this process, again choosing the two nodes with smallest distance, merging them, and so on until it has merged all the nodes into a single tree.

Let's look at how this works on our three species example, focusing first on the big picture and then returning to the Python details later.

Iteration 1. Our initial input is the matrix shown above and in Figure 11.1. The algorithm's first step is to find the pair of species with the minimum distance between them. In this example that pair is A and B with a distance of 6. Because of the molecular clock assumption, we conclude that these two are the most closely related. We'll therefore put them on the tree first.

Since the time that these two species diverged, 6 units of sequence distance have accumulated between them. We assume this change was clock-like, and thus an equal amount must have accumulated on each lineage. As 6/2 is 3, we can conclude that the divergence of A and B occurred 3 "sequence change" time units ago, and label our tree with this timing. Our partial tree containing A and B can be seen in Figure 11.1.

Before the first iteration completes, we need to update the distance matrix. By putting A and B on the tree, we have effectively merged them into a single new node. In later iterations, we'll treat this node like a species, potentially

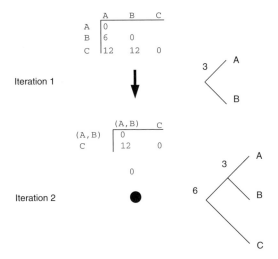

Figure 11.1. The UPGMA algorithm working on a simple input matrix.

combining it with other nodes. Therefore, we now create a new data matrix. Our new matrix will no longer have any rows or columns for A or B individually. Instead, it will have data for a combined "merged species" node (A, B) along with rows and columns for the other species from the original matrix, in this case just species C. In order to make this new matrix, we need to know what the distance should be between the new "merged species" (A, B) and species C. In the original distance matrix, A had a distance of 12 from C. B also had a distance of 12 from C. This makes our job easy: The "merged species" (A, B) should also have a distance of 12 from C. Our new matrix can be seen in Figure 11.1. You might be wondering what would have happened if the distance from A to C and the distance from B to C were not the same. We'll return to this question below.

Iteration 2. The data matrix now has just two species in it, the "merged species" (A, B) and species C. So, the closest pair is just (A, B) and C. We already have (A, B) on our tree and we'll now add C. The distance between these two nodes in the matrix is 12, which implies that the two lineages diverged 6 "sequence divergence" time units ago. We label our tree accordingly (Figure 11.1).

Now we're done; we've merged all the nodes into a single tree. The resulting tree contains information about the relationships between these three species and about the timing of their divergences.

Next let's consider another example using a four species matrix (solution illustrated in Figure 11.2).

	A	B	C	D
A	0			
B	24	0		
C	18	12	0	
D	24	4	12	0

Iteration 1. We find the nearest two nodes, which are B and D. Their distance is 4, implying they diverged 2 time units ago. We add them to the tree, labeling their divergence time.

We next create a new matrix with (B, D) as a node. In the starting matrix, B and D were each 24 distant from A. Thus (B, D) is also distance 24 from

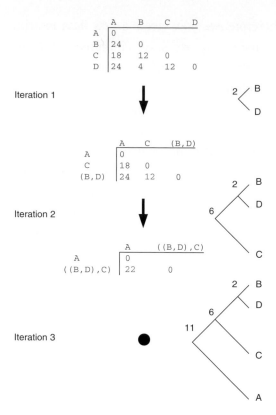

Figure 11.2. Another example of the UPGMA algorithm.

A. Similarly, B and D were each 12 distant from C, and so the distance from(B, D) to C is 12.

Iteration 2. The nearest two nodes are (B, D) and C. They diverged 12/2 or 6 time units ago. We add these to the tree.

We create a new matrix in which (B, D) and C are replaced by a single merged entry ((B, D), C) and we must now set the distance between((B, D), C) and A.

In this case, the distance from (B, D) to A (24) is different than the distance from C to A (18). To calculate the distance from ((B, D), C) to A we need to take a **weighted** average of the distances from (B, D) to A and C to A. The weight depends on the number of leaves in the two nodes being merged. (B, D) has two leaf nodes, and C has only one. So the distance from (B, D) to A should get twice the weight of the distance from C to A. Our calculation will be 2/3 × 24 + 1/3 × 18, which is 22.

Iteration 3. We finally merge ((B, D), C) and A, placing their divergence 22/2 or 11 time units ago. We're now down to a single node, and we stop.

11.2 Implementing UPGMA in Python

We've given a paper-and-pencil description of UPGMA, but now it's time to think about implementing the algorithm in Python. Our input will be a list of species and a distance matrix between those species.

We'll illustrate these with our three species example. You might imagine that the list of species would simply be a list like `['A', B', 'C']`. But, because we're trying to build a tree, and trees are tuples of the form `(Root, Subtree1, Subree2)`, it will actually be a bit easier if each species is itself a small tree such as `('A', (), ())` rather than just the string `'A'`. So, here's the species list in our example:

```
SpeciesList = [ ('A', (), ()),
                ('B', (), ()),
                ('C', (), ())  ]
```

While there are many different ways to represent a matrix, a dictionary representation is particularly convenient and easy to use. Recall that a dictionary comprises a set of key-value pairs. In our case, a key will be a pair of species and its value will be the distance between those two species. The distance between species `'A'` and species `'B'` in our example is 6. Since `'A'` is represented by the tree `('A', (), ())` and `'B'` is represented by the tree `('B', (), ())`, the key for this pair would be `(('A', (), ()), ('B', (), ()))`. So, our `Distances` matrix will look like this:

```
Distances = { (('A', (), ()), ('A', (), ())): 0,
              (('B', (), ()), ('B', (), ())): 0,
              (('C', (), ()), ('C', (), ())): 0,
              (('A', (), ()), ('B', (), ())): 6,
              (('B', (), ()), ('A', (), ())): 6,
              (('A', (), ()), ('C', (), ())): 12,
              (('C', (), ()), ('A', (), ())): 12,
              (('B', (), ()), ('C', (), ())): 12,
              (('C', (), ()), ('B', (), ())): 12  }
```

Note that here we include an entry for the distance from A to B and also from B to A.

Our algorithm will do the following for each iteration:

- Find the two nodes that are closest, considering all pairs of nodes in the `SpeciesList`.
- Remove these two nodes from the `SpeciesList`.
- Join these two to make a new node, and add it to `SpeciesList`.
- Update the distance matrix.

For the example data above, the closest pair we find initially is `('A', (), ())` and `('B', (), ())`. We remove these from `SpeciesList`, so that `SpeciesList` is now: `[('C', (), ())]`. Next we create a new node from our two closest species: `(3, ('A', (), ()), ('B', (), ()))`. Note that we've recorded the distance between these two species (3) in the root position of this node. We add it to `SpeciesList`, which now looks like this: `[('C', (), ()), (3, ('A', (), ()), ('B', (), ()))]`. Finally, we update the `Distances` matrix:

```
{      (('A', (), ()), ('A', (), ())): 0,
       (('B', (), ()), ('B', (), ())): 0,
       (('C', (), ()), ('C', (), ())): 0,
       (('A', (), ()), ('B', (), ())): 6,
       (('B', (), ()), ('A', (), ())): 6,
       (('A', (), ()), ('C', (), ())): 12,
       (('C', (), ()), ('A', (), ())): 12,
       (('B', (), ()), ('C', (), ())): 12,
       (('C', (), ()), ('B', (), ())): 12,
       (('C', (), ()), (3, ('A', (), ()), ('B', (), ()))): 12.0,
       ((3, ('A', (), ()), ('B', (), ())), ('C', (), ())): 12.0      }
```

The last two entries in red represent what has been added. Note that in our implementation, it's not actually necessary to delete anything from the dictionary representing the matrix. All you need to do is remove newly merged nodes from the `SpeciesList`.

Although you don't need to remove anything from the `Distances` matrix, you will need to remove nodes from the `SpeciesList`. Removing an item from a list can be done using Python's built-in `remove`. Here's an example:

```
>>> L = [2, 5, 9]
>>> L.remove(5)
>>> L
[2, 9]
```

11.3 Calibrating Trees

As we mentioned above, the molecular clock assumption allows UPGMA to put times on the various divergences in the tree that it builds. However, the "times" it produces are in units of sequence divergence. We can convert these to units of actual time if we have a known date for calibration. This might be the case, for example, if we have fossil evidence for the timing of one of the divergences.

Figure 11.3 shows an example of a tree constructed by UPGMA that we wish to calibrate. The divergences are given in units of sequence change. In this case,

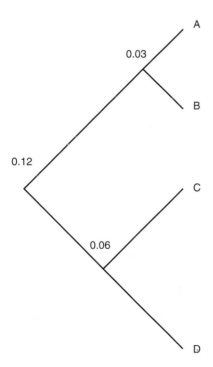

Figure 11.3. A tree calculated by UPGMA with dates in units of sequence change.

they represent substitutions (changes) per site, which is also what you'll see in the homework. Imagine that based on fossil evidence we believe that the divergence of species C and D occurred 1.2 million years ago. To calculate dates for the other nodes, we do the following. We divide 1.2 million years by 0.06 units of sequence change, which gives us 20 million years / unit sequence change. This represents the rate of the molecular clock. Using this value, we can estimate dates of the other divergences. The divergence for A and B will be 20 million years / unit sequence change × 0.03 units of sequence change or 0.6 million years. And the root divergence will be 20 million years / unit sequence change × 0.12 units of sequence change or 2.4 million years.

We'll use this technique in the homework to date divergences in human evolution.

Putting it All Together

The UPGMA algorithm that you've learned in this chapter is one of many phylogenetic reconstruction methods in current use. One of its attractive features is that it is fast: It can construct very large phylogenetic trees very quickly. On the other hand, it is also known to construct incorrect phylogenetic trees in some circumstances. Therefore, other algorithms have been developed that are similar in structure to UPGMA but use more biologically accurate methods to join species into trees and join small trees into larger ones. One such method is called Neighbor Joining. Other approaches use biological sequence data directly, rather than the distance matrices that UPGMA and Neighbor Joining use. One such approach is based on the maximum likelihood method, which we'll explore in a different context in Chapter 13. Though phylogenetic reconstruction was one of the first areas of computational biology, it continues to be an area of active research.

In the final programming problem for Part III, you'll use the tools we've developed to create and display phylogenetic trees based on mitochondrial DNA from modern humans and a Neanderthal. Your trees will tell you something about the relationship between these species, and allow you to distinguish between alternative models of modern human origins.

FOR FURTHER READING

Campbell BG, Loy JD and Cruz-Uribe K. *Humankind Emerging*, 9th edn. Pearson, 2005.

Green RE, Krause J, Briggs AW, et al. A draft sequence of the Neandertal genome. *Science* 2010;328 (5979):710–722.

Krause J, Briggs AW, Kircher M, et al. A complete mtDNA genome of an early modern human from Kostenki, Russia. *Current Biology* 2010;20(3):231–236.

Green RE, Malaspinas AS, Krause J, et al. A complete Neandertal mitochondrial genome sequence determined by high-throughput sequencing. *Cell* 2008;134(3):416–426.

Ingman M, Kaessmann H, Pääbo S and Gyllensten U. Mitochondrial genome variation and the origin of modern humans. *Nature* 2000;408:708–713.

PART IV

Additional Topics

. . . .

This unit contains three chapters on three different topics: RNA folding, gene regulation, and genetic algorithms. Each chapter presents a problem that leads to a programming project. Each chapter also introduces some major new ideas that are transferable to other areas of computer science and computational biology. For example, Chapter 12 examines the use of more sophisticated recursion in the context of RNA folding, Chapter 13 introduces the method of maximum likelihood in the context of inferring gene regulatory networks, and Chapter 14 introduces the concepts of algorithm efficiency, NP-hardness, and genetic algorithms.

RNA Secondary Structure Prediction

> **Computing topics:**
> recursion,
> memoization

Acquired Immune Deficiency Syndrome or AIDS was first recognized in the early 1980s. Within a few years of its discovery, scientists had identified the virus that causes it, called the Human Immunodeficiency Virus or HIV. Through the 1980s and early 1990s, the HIV/AIDS epidemic grew steadily, with the number of affected people and the number of deaths increasing every year. By the late 1990s it was causing millions of deaths per year worldwide, and had a prevalence of more than 20% in the adult population of some countries.

An epidemic of this magnitude requires multiple types of response. One approach has been to limit the spread of the disease through education, for example, by encouraging the use of condoms and discouraging drug users from sharing needles. A second approach has been to carry out research into the basic biology of HIV in an effort to develop better treatments or even a vaccine.

The programming problems for this section are connected with basic research into HIV. They involve predicting the RNA secondary structure in a single gene from the HIV genome. We'll talk about secondary structure and its prediction shortly. But first, let's briefly discuss HIV's genome and life cycle.

Most organisms that you're familiar with have genomes made of DNA. However, some viruses use RNA as their genetic material instead of DNA. HIV comes from a group of viruses that have single-stranded RNA genomes. Interestingly though, the life cycles of these viruses do involve a double-stranded DNA stage. When an HIV viral particle infects a human cell, it injects its RNA genome as well as several proteins into that cell. Among the proteins injected is one called *reverse transcriptase*, whose function is to copy the RNA genome into double-stranded DNA. This double-stranded DNA can then be inserted into the host cell's genome. HIV is called a *retrovirus* because this process of copying genomic information from RNA into DNA is the reverse of the typical pattern.

Once it has inserted itself into a host cell's genome, HIV can do one of several things. It can enter a period of latency, where it quietly sits in the genome, evading detection by the immune system. Another possibility is that it can become actively transcribed, ultimately leading to the production of new viral particles. In the case of transcription, the host DNA embedded HIV serves as a template for new RNA genomes. These become templates themselves for translating HIV proteins. Ultimately both RNA genomes and proteins are packaged into new viral particles.

Our homework problems in this section focus on the Pol gene of HIV, which codes for several proteins including reverse transcriptase. There's a region within this gene where the RNA can fold into a three-dimensional structure that is crucial for the proper translation of the gene. In the homework problems we'll be seeking to identify and visualize this region. To that end, we'll develop an algorithm to predict RNA secondary structure. We'll first describe secondary structure and put it in the broader context of RNA folding.

RNA is a molecule with diverse functions in biology. Many of those functions depend on its ability to fold into particular three-dimensional shapes. As in proteins, we speak of several levels of structure in RNA. The linear sequence of nucleotides is referred to as the *primary structure*. This linear sequence can fold into particular shapes, a process that involves a number of different kinds of chemical interactions. A first category consists of base-pairing interactions, where nucleotides form hydrogen bonds with specific other nucleotides (e.g., A pairing with U). The structure that results from this process is called *secondary structure*. Figure 12.1 shows an example.

After secondary structure forms, additional interactions arise between different parts of the molecule. These additional interactions are called *tertiary structure* interactions, and they cause the molecule to fold up into its final three-dimensional shape. Tertiary structure interactions do not involve base pairings, but can involve other kinds of hydrogen bonds.

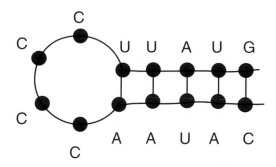

Figure 12.1. An RNA secondary structure motif called a hairpin loop.

We'll develop a method for predicting RNA secondary structure computation-ally. Of course, the secondary structure by itself does not tell us the final folded shape of an RNA molecule. However, predicting secondary structure is much easier than predicting the full three-dimensional shape, and it turns out that knowing the secondary structure frequently provides crucial information about the function of an RNA molecule.

"Industrial strength" RNA secondary structure prediction algorithms are based on the thermodynamics of the folding process. These algorithms calculate the energy of the molecule in various folded states and try to find the minimum energy state. Such algorithms are complicated, and here we'll use a slightly simpler one, called *Nussinov's algorithm*, that involves maximizing the number of base-pair interactions formed. We'll allow three types of pairs: A-U, C-G, and G-U.[1] Our algorithm will award one point for a pair, and zero points for unpaired bases. The goal is to return the maximum score possible for a given input RNA sequence.

We'll assume that an individual base can form at most one pair – once a base has paired it becomes inert and cannot pair again. And, initially, we'll also assume that a base can pair with any other matching base on the strand. That is, an A can pair with any U, even one right next to it. In the homework we'll revisit this assumption and modify our algorithm to be slightly more realistic.

So, to get started, it will be handy to have a function that determines whether two bases can pair. This is easy and here we have it:

---

[1] The G-U pair represents a non-standard type that sometimes occurs in RNA.

```
def isComplement( base1, base2 ):
    ''' Returns boolean indicating if 2 RNA bases are
    complementary.'''
    if base1=='A' and base2=='U':
        return True
    elif base1=='U' and base2=='A':
        return True
    elif base1=='C' and base2=='G':
        return True
    elif base1=='G' and base2=='C':
        return True
    if base1=='G' and base2=='U':
        return True
    elif base1=='U' and base2=='G':
        return True
    else:
        return False
```

Next, we'll make a simplifying assumption that is used in a number of RNA folding algorithms. This is called the "no pseudo-knots" assumption and here's what it says. When two nucleotides pair to each other, they form a loop. Bases inside that loop can pair with one another but they cannot pair with bases outside the loop. And outside this loop, bases can pair with one another, but not with bases inside the loop. This is illustrated in Figure 12.2. The RNA in 12.2(a) folds without pseudo-knots whereas the one in 12.2(b) has a pseudo-knot indicated in red.

Here's another way to look at this. If we draw the RNA out in a line as in Figure 12.2(c) and (d), the connections between paired bases cannot cross one another. So, the folding in Figure 12.2(c) has no pseudo-knots while the one in Figure 12.2(d) has a pseudo-knot, again indicated in red.

While pseudo-knots can arise in RNA folding, they are relatively uncommon. Thus, eliminating them from consideration will give us reasonably good predictions of RNA folding and lends itself to a nice computational solution.

Now, let's write our RNA folding function, which we'll call fold(RNA). It takes an RNA strand as input, and returns the maximum score, that is, the maximum number of base pairs it can form with itself. We begin with the base case. A natural

Illustration 1: RNA drawn as folded

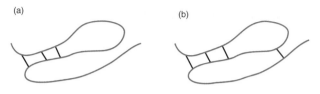

(a)   (b)

Illustration 2: RNA drawn linearly

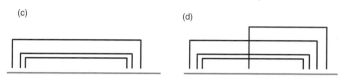

(c)   (d)

Figure 12.2. Illustration of pseudo-knots using two methods of drawing. (a) An RNA strand folded into a hairpin with no pseudo-knots. Base-pairing interactions shown in black. (b) The same fold, but this time an additional base-pairing interaction (red) has formed between a base in the loop and one outside. This is a pseudo-knot. (c) Same scenario, but with a linear representation of the RNA. Base-pairing interactions are shown as black upside down U's. (d) In this representation a pseudo-knot is a line (red) which crosses the other lines.

base case is when the input string is less than two in length. In this case, there can be no pairs, and we should return 0.

```
def fold( RNA ):
    ''' Takes an RNA sequence as input and returns the maximum number
        of base pairs it can form with itself.'''
    if len(RNA)<2:
        return 0
```

Next, we deal with the general case. Consider the following example RNA sequence: ACCCACCCUCCCUCC. This string's length is not less than 2, and so we go to the general case.

In the general case, let's consider the first base in our string. In this example, it's an A. There are only two possibilities: That leading A is paired with a complementary base (a U) later in the sequence or it's not paired at all. This sounds like the use-it-or-lose-it idea – and indeed it is!

We begin with the lose-it case, where the A is "lost" by being unpaired. In this case, the score of the optimal solution will simply be the score of fold on the RNA

```
RNA = ACCCACCCUCCCUCC
```

Figure 12.3. An example RNA sequence.

with the first base removed `fold(RNA[1:])`. Thus our function to this point should look like this:

```
def fold( RNA ):
    ''' Takes an RNA sequence as input and returns the maximum number
    of base pairs it can form with itself.'''
    if len(RNA)<2:
        return 0
    else:
        bestSoFar = fold(RNA[1:])   # lose-it case
```

Next, we consider the use-it case where the A is paired with a U. The challenge here is that we can't tell yet which U should pair with the leading A to get the best score. In our example (Figure 12.3), there are two possible U's that could match this A. But, in general, there might be no possible matches or there might be many. We're going to need to consider all of them.

So, we'll use a `for` loop to consider each base in the sequence. For each base, we'll check to see whether it can pair with our first base, the A in this example. And if it can pair, we'll use recursion to consider that option.

More specifically, our `for` loop will start with the base at position 1 and go to the last index in the string. At each position we'll call the `isComplement` on our first base (A in the example) and the base at that position. If `isComplement` returns `False`, we move on to the next position. But if it returns `True`, then we have a candidate pair, and we somehow need to calculate the score in the case of this pairing. We'll get to that last step in a moment, but here's what we have so far:

```
def fold( RNA ):
    ''' Takes an RNA sequence as input and returns the maximum number
        of base pairs it can form with itself.'''
```

```
if len(RNA)<2:
    return 0
else:
    bestSoFar = fold(RNA[1:])    # lose-it case
    for i in range(1,len(RNA)): # use-it cases
        if isComplement(RNA[0],RNA[i]):
```

What do we do when we find a complementary base? Let's consider our example RNA sequence again (Figure 12.3) and, for the sake of example, let's look at the case where our `for` loop is now at position 8. This position has a U that can pair with the A at position 0. When our `for` loop gets to position 8, `isComplement` will return `True` and we'll find ourselves inside the `if` block. Our task here is to compute the best score for the case that the leading A matches with this U. This might not ultimately be the best overall score for this RNA string, but it's a candidate that we need to consider.

The score for this case will consist of three parts. First, we'll get 1 point for the pairing between the A and the U. Second, because of the no pseudo-knot assumption, we know that the stretch of bases in positions 1 through 7 (that is, between the A and the U that we're considering) can pair with one another, but not with a base to the right of that U. So, the stretch from position 1 to 7 is a separate RNA folding problem. We can recursively call `fold(RNA[1:7])` to determine the maximum number of points for that region. Third, the bases in positions 9 and beyond can pair up among one another, but not with the bases in positions 1 to 7. So, this is another separate RNA folding problem. We can call `fold(RNA[9:])` to find the maximum number of points for this stretch of RNA. So, the best score for the entire RNA for the case that the leading A pairs with the U at position 8 is:

```
1 + fold(RNA[1:8]) + fold(RNA[9:]
```

But we've already noted that this might or might not be the best score for the entire RNA string. It might be that when the leading A pairs with a different U, we get a better score, or that the aforementioned lose-it option is the best one.

So, after computing each candidate score, we'll compare it to the best score we've found so far. If the newly computed score is better than the best one we've seen so far, we'll update our best score so far. The code for the whole `fold` function looks like this:

```
def fold( RNA ):
    ''' Takes an RNA sequence as input and returns the maximum number
        of base pairs it can form with itself.'''
    if len(RNA)<2:
        return 0
    else:
        bestSoFar = fold(RNA[1:])     # lose-it case
        for i in range(1,len(RNA)): # use-it cases
            if isComplement(RNA[0],RNA[i]):
                score = 1 + fold(RNA[1:i]) + fold(RNA[(i+1):])
                if score > bestSoFar:
                    bestSoFar = score
        return bestSoFar
```

Let's briefly return to the importance of the no pseudo-knots assumption. Consider our example above where the `for` loop had reached position 8. We then made two separate recursive calls: One on the inside (between the A and U) and one on the outside (to the right of the U). Because we don't permit pseudo-knots, we know that those two subproblems can be solved independently: If this A and U are matched, there cannot be any pairing between a base on the inside and a base on the outside. If pseudo-knots were permitted, we could not have made these two independent recursive calls.

In general, recursion requires that the subproblems that it solves are independent of one another. Computer scientists have a fancy name for this: It's called the *independent substructure property*.

The no pseudo-knots assumption gives us that property for the RNA folding problem.

In the programming problems for this chapter, you'll implement a faster version of the Nussinov algorithm we described here. To do this you'll use the memoization technique from Chapter 7. After that, you'll further extend your program to compute the actual pairings of the nucleotides and draw the folded RNA. Finally, you'll use your software to explore the predicted folds in the HIV Pol gene. In general, knowing the folding patterns of an RNA sequence can help us determine its function.

Since the Nussinov algorithm was first proposed, researchers have developed increasingly more complex alternatives. Like Nussinov, these algorithms are recursive and rely on memoization (or the related dynamic programming technique) to run fast. However, they also incorporate more sophisticated thermodynamic models of the folding process to get even more accurate predictions. One such algorithm is the Zuker algorithm, which is implemented in the mfold web application (http://mfold.rna.albany.edu). You can try it out and see how its solutions compare to the ones that your program computes.

FOR FURTHER READING

Eddy SR. How do RNA folding algorithms work? *Nature Biotechnology* 2004;22(11):1457–1458. http://www.nature.com/nbt/journal/v22/n11/abs/nbt1104–1457.html.

Nussinov R and Jacobson AB. Fast algorithm for predicting the secondary structure of single-stranded RNA. *Proceedings of the National Academy of Sciences* 1980;77(11):6309–6313. http://www.pnas.org/content/77/11/6309.short.

Zuker M and Stiegler P. Optimal computer folding of large RNA sequences using thermodynamics and auxiliary information. *Nucleic Acids Research* 1981;9(1):133–148. http://www.ncbi.nlm.nih.gov/pmc/articles/PMC326673/.

13 Gene Regulatory Networks and the Maximum Likelihood Method

> **Computing topics:**
> the maximum likelihood method

The different cell types in the human body look different and do very different things: Compare, for example, liver cells and brain cells. How do they manage to be so different given that they have the same DNA? The answer is that cells regulate the *expression* of their genes – that is, they control when and where their genes are used to make protein. As a result, different cell types make a different complement of proteins.

In fact, the expression of a gene can be regulated by other genes. Biologists represent this using a *gene regulatory network*, a diagram that shows how genes interact. Figure 13.1 shows an example of such a network for some genes in the bacterium *Bacillus subtilis*. In the diagram, each gene is represented by a circular *node*. To show that one gene regulates another, we draw an *edge*, that is, a line with an arrow. This indicates that one gene (the one which the arrow is drawn from) regulates the transcription of the second (the one which the arrow is drawn to). The effect of this regulation might either be positive (*upregulation*) or negative (*downregulation*), but we won't make a distinction between those two cases here.

Inferring gene regulatory networks contributes to an overall understanding of cell physiology and is important for many applications such as the development of therapies for diseases. In this chapter, you'll learn a powerful technique for discovering gene regulatory networks and you'll implement it in Python. The

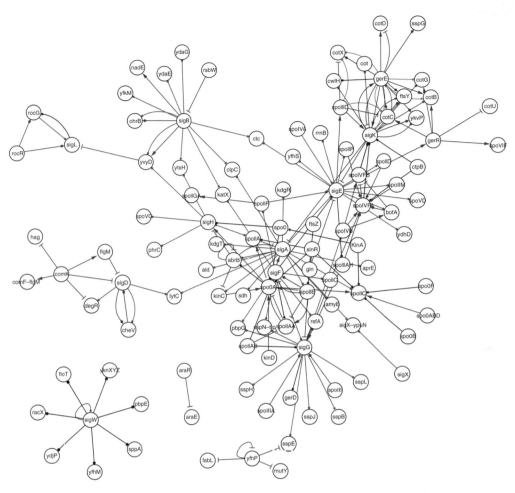

Figure 13.1. A gene regulatory network (source: http://2013.bionlp-st.org/tasks/gene-regulation-network).

method is called *maximum likelihood estimation* and it's used in many applications in computational biology and throughout the sciences.

Imagine that we have a set of genes whose expression we are interested in studying.[1] We'd like to know how their expression changes under different experimental conditions. The conditions represent various kinds of manipulations which the experimenter performs before measuring expression – for example, treatments

---

[1] The experimental measurement of gene expression is a major focus of biological research. There are many ways of doing such measurements, and new methods are constantly being developed.

with different chemical compounds. We measure the expression level of each gene under each condition. For simplicity, we'll assume that the expression level is either 0 (low) or 1 (high). In practice, the levels of expression could take on other intermediate values as well.

Let's consider an example where we have two genes named gene 0 and gene 1 and four experimental conditions. In this case, the expression profiles might look like this:

```
gene 0:   0001
gene 1:   1001
```

Each row corresponds to a gene and each column corresponds to a particular condition (e.g., treatment with a different compound). In the condition in the first column, gene 0 happens to have low expression and gene 1 has high expression.

Figure 13.2 shows the four possible regulatory networks that can explain these data. In Figure 13.2(a), there are no edges, meaning that neither of the two genes regulates the other. Figure 13.2(b) has an edge from gene 0 to gene 1, indicating that gene 0 regulates gene 1. Figure 13.2(c) has an edge from gene 1 to gene 0, indicating that gene 1 regulates gene 0, and Figure 13.2(d) has edges in both directions, indicating that each gene regulates the other. For reasons that will become evident soon, networks like Figure 13.2(d) are problematic because they have cycles. We'll ignore networks with cycles in this chapter, although you'll notice that the regulatory network in Figure 13.1 actually has some cycles.

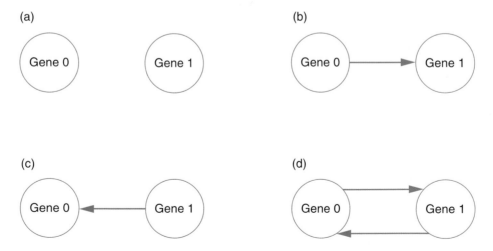

Figure 13.2. Four possible regulatory networks for two genes.

These four networks are simply four different hypotheses of how our two genes interact. Which of the three cycle-free (or *acyclic*) networks in Figure 13.2 is the right one? While knowing what's absolutely "right" is difficult, we can use computational methods to infer which one is most likely given the expression data that we've observed.

The maximum likelihood method considers each candidate gene regulatory network one-by-one. Each network is a hypothesis for the actual network. For each such hypothesis we ask, "what's the probability that we would have observed our expression data if this was the network that regulates the genes?" We compute that probability for each network and then choose the network that gives us the maximum probability. In general, the *likelihood* is the probability of observing a given set of data assuming a particular hypothesis. The maximum likelihood method reports the hypothesis that maximizes this likelihood.

Recall that our expression data are:

```
gene 0:   0001
gene 1:   1001
```

Let's now compute the likelihood of these expression data for the gene regulatory network in Figure 13.2(a), in which neither gene regulates the other. We'll introduce our technique with a non-biology-related example. Imagine that you watched a friend flipping a "magic" coin 100 times. In 75 of those 100 flips, the coin landed on heads and in 25 it landed on tails. You could reasonably assume from your observations that the probability of the coin landing on heads is $75/100 = 3/4$ and the probability of it landing on tails is $25/100 = 1/4$.

Now that you've observed the coin being flipped and have estimated those probabilities, your friend says to you: "I'm going to flip my coin three times and if we first get a heads, then a tails, and then another tails, I'll buy you a pizza for dinner." What's the probability that you'll be getting a pizza? Based on your estimates of the probabilities of heads and tails, the probability of the sequence heads-tails-tails is $\frac{3}{4} \times \frac{1}{4} \times \frac{1}{4}$. You'll probably want to make alternate dinner plans!

We can do exactly the same thing with our gene expression data in the case of Figure 13.2(a). For gene 0 with expression profile 0001, we see that 0 occurs with frequency $\frac{3}{4}$ and 1 occurs with frequency $\frac{1}{4}$. We use those frequencies as our best estimates for the probabilities of 0's and 1's occurring, and from this we estimate that the probability of the expression profile 0001 for gene 0 is $\frac{3}{4} \times \frac{3}{4} \times \frac{3}{4} \times \frac{1}{4}$.

Similarly, for gene 1 with profile 1001, 0 occurs with frequency ½ and 1 occurs with frequency ½, so the probability of this pattern is ½ × ½ × ½ × ½. The probability of observing the pattern 0001 for the first gene and 1001 for the second gene is therefore the product of these two, which is ¾ × ¾ × ¾ × ¼ × ½ × ½ × ½ × ½ = 0.006592. That's the likelihood of observing these expression data if the two genes are independent as in the gene regulatory network in Figure 13.2(a).

Next, let's consider the likelihood of observing those data assuming the gene regulatory network in Figure 13.2(b), where gene 0 is assumed to regulate gene 1. The probability of observing 0001 for gene 0 is, as before, ¾ × ¾ × ¾ × ¼. But now, since gene 1 is assumed to be regulated by gene 0 in this scenario, the probability of observing the pattern 1001 for gene 1 is dependent on the observed pattern 0001 for gene 0. This is called a *conditional probability*.

Gene 1 has the expression profile 1001. The first column contains a 1. Since gene 0 is assumed to regulate gene 1 in our current hypothesis, we look at the value for gene 0 in that column: Its value is 0. So, now we ask, "what's the probability that gene 1 would have value 1 given that gene 0 has value 0?" To answer this question, we look at the number of columns where gene 0 has value 0. There are three such columns. Among those three columns, how many of them have a 1 for gene 1? Just one – the first column. So, the probability that gene 1 is a 1 given that gene 0 is a 0 is ⅓.

Next, we consider the second column for gene 1. Its value is 0. And, since we're assuming that gene 1 is regulated by gene 0, we look at the value for gene 0 in that column and see that it's also 0. So, we ask, "what's the probability of seeing 0 in gene 1 assuming 0 in gene 0?" There are three columns that contain 0's for gene 0 and, among those three columns, two have 0's for gene 1. So, the conditional probability is ⅔. We get the same value for the third column. Finally, for the last column, we see 1 in gene 1 and 1 in gene 0. So, we ask, "what's the probability of seeing 1 in gene 1 assuming the value 1 in gene 0?" Only one column has a 1 in gene 0, so the conditional probability is 1/1 = 1. In total, the likelihood for this network in Figure 13.2(b) is therefore the product of ¾ × ¾ × ¾ × ¼ for gene 0 and ⅓ × ⅔ × ⅔ × 1 for gene 1 conditioned on gene 0, for a total of 0.015625. That's approximately twice the likelihood for the network in Figure 13.2(a). That makes sense, because the two rows for genes 0 and 1 are almost identical, so we'd expect that those genes might regulate one another somehow. We encourage you to try doing the analogous computation for the network in Figure 13.2(c) and determine which of the networks has the highest likelihood.

Let's look at one slightly larger example before we move on to discussing how this analysis will be done in your computer program. Consider the expression data below for four genes numbered 0 through 3 and five conditions:

```
Gene 0:   01010
Gene 1:   11110
Gene 2:   00010
Gene 3:   11111
```

Our plan is to evaluate all of the possible gene regulatory networks for four genes, compute the likelihood of each one, and return the regulatory network of maximum likelihood. That's a big task, but let's see how the likelihood would be computed for just one network, using the example network in Figure 13.3.

The likelihood of this candidate network is the probability of the expression data for gene 2 multiplied by the probability of the observed data for gene 0 conditional on the data observed for gene 2 (because gene 2 regulates gene 0), multiplied by the probability of the observed data for gene 1 conditioned on the data for both gene 2 and gene 0 (since they both regulate gene 1), multiplied by the probability of the data for gene 3 conditioned on the data for gene 1 (since gene 1 regulates gene 3).

Gene 2 is not regulated by any other gene, so we'll start by computing the probability of observing the expression data 00010 for gene 2 given this regulatory network. In that row, four of the five data points are 0 and one out of five is a 1. So, the probability of the sequence 00010 is simply $\frac{4}{5} \times \frac{4}{5} \times \frac{4}{5} \times \frac{1}{5} \times \frac{4}{5}$.

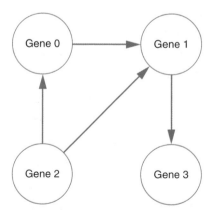

Figure 13.3. One gene regulatory network for four genes.

Gene 0 is only regulated by gene 2, so let's compute that one next. In the first column for gene 0, we see a 0. Gene 2 has a 0 in that position. What's the probability of seeing a 0 in gene 0 assuming a 0 in gene 2? Gene 2 has four 0's and in three of those four columns gene 0 has the value 0. So, the probability is ¾. We move on to the next column and see that gene 0 has a 1 there while gene 2 has a 0 there. Again, gene 2 has four 0's and only one of those four columns contains a 1 in gene 0's row, so the probability is ¼. We continue this way, multiplying those values to get the conditional probability for gene 0's data. Try it yourself! We get the following term for gene 0 conditioned on gene 2: ¾ × ¼ × ¾ × 1 × ¾.

Things get more interesting when we move on to gene 1. It is regulated by both genes 0 and 2 in our candidate network. We look at the first column for gene 1 and see that it's a 1. Gene 0 has a 0 in that column and gene 2 has a 0 in that column. So, now we need to compute the probability that gene 1 is a 1 when gene 0 is a 0 and gene 2 is a 0. We count the number of columns in which *both* gene 0 is a 0 and gene 2 is a 0. There are three such columns (columns 0, 2, and 4). And gene 1 has a 1 in two of those three columns. So, the conditional probability of gene 1 being a 0 given that gene 0 is a 0 *and* gene 2 is a 0 is ⅔.

We continue in the same spirit for the remaining columns for gene 1. We get the following: ⅔ × 1 × ⅔ × 1 × ⅓.

Finally, we move on to gene 3. For our current network in Figure 13.3, gene 3 is regulated by gene 1. The first column for gene 3 contains a 1 and gene 1 contains a 1 in that column too. Gene 1 has four 1's and in all four of those cases gene 3 contains a 1, contributing $4/4 = 1$ to the probability. In total, we get $1 × 1 × 1 × 1 × 1 = 1$ for gene 3 conditioned on gene 1. Thus, in total, the likelihood of observing these expression data conditioned on this network is the product:

$$(⅘ × ⅘ × ⅘ × ⅕ × ⅘) × (¾ × ¼ × ¾ × 1 × ¾) × (⅔ × 1 × ⅔ × 1 × ⅓) × (1 × 1 × 1 × 1 × 1) = 0.00128$$

Why did we not consider gene regulatory networks with cycles? Indeed, it's quite possible that a pair of genes regulates one another or there exists some more complex cycle of interactions. The problem with cycles is a computational one and not a biological one. We compute the probability of a gene's pattern conditioned on the patterns of the genes that regulate it in the network under consideration. If two genes regulate one another, then we have a chicken-and-egg problem. One gene's contribution to the likelihood is computed assuming the pattern for the other gene and vice versa. Addressing networks with cycles gets complicated and thus we've cut a corner. It's not uncommon for scientists to make compromises like this in order to find reasonably good solutions, even if they aren't always optimal.

With only four genes, we could enumerate all of the possible acyclic networks, compute the likelihood for each one, and report the network with maximum likelihood. However, as we have more genes interacting, the networks get large fast. For example, for ten genes there are 4175098976430598143 different acyclic networks to test. Even if we had a very powerful computer capable of evaluating one trillion networks per second, it would take 48 days to complete this task. And that's just for ten genes!

So, in practice we need some sort of *heuristic*, a fast way of getting a good, but not necessarily optimal, solution. When you implement this problem, your main function will simply generate some user-specified number of random acyclic networks where the number of nodes is equal to the number of genes. It will then find the likelihood of each one and return the network with maximum likelihood. More sophisticated heuristics are also possible.

How can we generate a random acyclic network? One way to do it is to first generate a totally random network and then check to see whether it has a cycle. That works, but it's effortful. A simpler approach is to only build networks that are guaranteed to be acyclic as follows. Imagine that we have four genes and thus need a network with four nodes numbered 0 through 3. We'll first scramble, or *permute*, those nodes. Fortunately, Python's random module has a function called shuffle that does this.

Say that we now have a list of permuted nodes [2, 0, 1, 3]. Now, we'll add the directed edges. If we make sure that a node only has edges to nodes later in the list, we can't have any cycles. Do you see why? So, node 2 gets some random subset of nodes 0, 1, and 3 as its "neighbors." Let's say that our randomized process gives node 2 neighbors 0 and 1. We can represent this by the list [2, 0, 1] indicating that there are edges *from* node 2 *to* nodes 0 and 1 (or in other words, 2 regulates 0 and 1). Moving on to node 0, we choose some subset of 1 and 3 as its neighbors. Say we choose just 1. We represent this with the list [0, 1]. Now, we move onto node 1, which either has no neighbor or just neighbor 3. Let's assume it gets neighbor 3. We represent this as [1, 3]. Finally, node 3 can have no neighbors because there are no other nodes after it in the permuted list. So, our final random network in this case is:

```
[ [2, 0, 1], [0, 1], [1, 3], [3] ]
```

Just by "luck," this turns out to be the network in Figure 13.3! This random generation process guarantees that the created graph has no cycles. Just as important, every graph with no cycles has a chance of getting created by this process. If that were not the case, we might miss the opportunity to generate the best possible solution.

Once we have a network, we're ready to compute the likelihood of our given expression data given that network. The expression data are a matrix and can be represented as a list of lists, where each constituent list represents the row for one gene. For example, the expression data...

```
Gene 0:   01010
Gene 1:   11110
Gene 2:   00010
Gene 3:   11111
```

... would be represented by the list

```
A = [ [0, 1, 0, 1, 0],
      [1, 1, 1, 1, 0],
      [0, 0, 0, 1, 0],
      [1, 1, 1, 1, 1] ]
```

Notice that A[0] gives us the first row and thus A[0][4] gives us the element at index 4 in that row, which is the upper-right 0 in the matrix.

Next, we'll write a function likelihood(expressionData, network) that computes the likelihood of the given expressionData with respect to the given network. Consider the network that we've been using in our examples so far: [[2, 0, 1], [0, 1], [1, 3], [3]]. We consider each of the lists in this network in turn. When we look at the list [2, 0, 1], for example, we're computing the probability for node 2. When we look at the list [0, 1] we're computing the probability for node 0, and so forth. Let's take a closer look at the computation when we get to the list [1, 3] and we're computing the probability for node 1.

To compute the probability for node 1, we need to first find which nodes regulate node 1. Those are the nodes that have edges *to* node 1. We'll need a short helper function to find those nodes. Notice that the list [2, 0, 1] and the list [0, 1] both contain node 1, so nodes 2 and 0 have edges to node 1. (Of course, the list [1, 3] contains node 1, but that's not interesting because that list begins with 1, indicating that it's the list of nodes with edges *from* node 1.) So, our helper function will return the list [2, 0] when we ask it who regulates node 1.

Finally, we use this information to compute the probability for node 1. We look at the expression profile for node 1 (gene 1), column-by-column. In the first column we see a 1. Since gene 1 is regulated by genes 0 and 2, we look at the entries for genes 0 and 2 in that column. They are 0 and 0. Now, we look at each column that contains a 0 and 0 for genes 0 and 2, respectively, counting the number of those columns that contain a 1 and dividing it by the total number of such columns. There are two such columns that contain a 1 in gene 1 and there are three such columns in total, resulting in a contribution of ⅔ to our total product of probabilities.

Putting it All Together

The study of gene regulatory networks is of great importance both in understanding the normal function of organisms and in combating disease. Defects in gene regulation and regulatory networks are implicated in many diseases, including cancer.

Here we have used the maximum likelihood method to infer the most likely gene regulatory networks for a given set of conditions. This method is also used to solve a great variety of other biological problems, including inferring phylogenetic trees, inferring the structures of macromolecules, and many others.

While we made a number of simplifying assumptions in our model (e.g., assuming that the gene expression was binary rather than a continuous set of values between 0 and 1, that the networks were acyclic, etc.), more complex models and algorithms for inferring gene regulatory networks continue to be studied.

In the programming problem for this chapter you'll implement the method that we described here, allowing you to infer and draw the maximum likelihood gene regulatory network.

14 Birds, Bees, and Genetic Algorithms

> **Computing content:**
> algorithm efficiency,
> computational complexity,
> genetic algorithms

This chapter is about the birds and the bees, but it's probably not what you're thinking! Let's start with the backstory.

Imagine that you're a salesperson who needs to travel to a set of cities to show your products to potential customers. The good news is that there's a direct flight between every pair of cities and, for each pair, you're given the cost of flying between those two cities. Your objective is to start in your home city, visit each city *exactly once*, and return back home at lowest total cost. This is called the *Traveling Salesperson Problem* and it's one of the most famous problems in computer science.

For example, consider the set of cities and flights shown in Figure 14.1 and imagine that your start city is Aville.

A tempting approach to solving the Traveling Salesperson Problem is to use an approach like this. Starting at our home city, Aville, fly on the cheapest flight. That's the flight of cost 1 to Beesburg. (This is not yet the part about the bees.) From Beesburg, we could fly on the least expensive flight to a city that we have not yet visited, in this case Ceefield. From Ceefield, we would then fly on the cheapest flight to a city that we have not yet visited. (Remember, the problem stipulates that you only fly to a city once, presumably because you're busy and you don't want to fly to any city more than once – even if it might be cheaper to do so.) So now, we fly from Ceefield to Deesdale and from there to Eetown. Uh oh! Now, the constraint

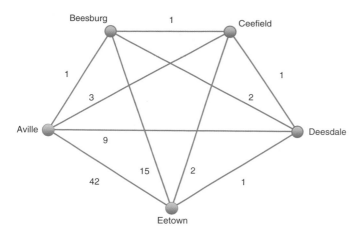

Figure 14.1. Cities and flight costs.

that we don't fly to a city twice means that we are forced to fly from Eetown to Aville at a cost of 42. The total cost of this "tour" of the cities is $1 + 1 + 1 + 1 + 42 = 46$.

This is a *greedy algorithm* because at each step it tries to do what looks best at the moment, without considering the long-term implications of that decision. This greedy algorithm didn't do so well here. For example, a much better solution goes from Aville to Beesburg to Deesdale to Eetown to Ceefield to Aville with a total cost of $1 + 2 + 1 + 2 + 3 = 9$. In general, greedy algorithms are fast but, as we saw in Chapter 6, they often fail to find optimal or even particularly good solutions.

It turns out that many animals confront a similar problem. For example, Clark's Nutcrackers have been shown to find remarkably good tours as they collect the thousands of pine nuts that they bury for winter.[1] And, it's not just birds. A bee tours a set of flowers in search of nectar and it would like to minimize its time and energy by finding a shortest tour. While the Traveling Salesperson Problem is notoriously difficult for humans and even computers, some birds and bees are evidently quite good at it.[2]

[1] Kamil AC and Jones JE. The seed-storing corvid Clark's nutcrackers learn geometric relationships around landmarks. Nature 1997;390:276–279.

[2] Lihoreau M, Chittka L and Raine NE. Travel optimization by foraging bumblebees through readjustments of traplines after discovery of new feeding locations. *The American Naturalist* 2010;176(6): 744–757.

This chapter is about the "hardness," or *computational complexity*, of problems. At the end of this chapter, you'll use a technique called *genetic algorithms* to find good – although not necessarily optimal – solutions to the Traveling Salesperson Problem.

Throughout this book, we've explored a number of algorithms for solving problems of interest to biologists. For example, we explored algorithms for finding genes, aligning sequences, inferring phylogenies, and folding RNA. Most of these algorithms were quite fast, although some were faster than others. But for some problems, like the Traveling Salesperson Problem, no fast algorithms are known.

Roughly speaking, computer scientists categorize computational problems into two broad categories: Those that can be solved by fast algorithms and those for which the only known algorithms are slow. We need to be more precise about "fast" and "slow" and that's what we'll do next.

14.1 Fast Algorithms

Consider the problem of taking a list of n numbers and finding the smallest number in that list. One simple algorithm simply "cruises" through the list, keeping track of the smallest number seen so far. We'll end up looking at each number in the list once, and thus the number of steps required to find the smallest number is roughly n. A computer scientist would say that the running time of this algorithm "grows proportionally to n."

Imagine that we're given a list of n numbers and we wish to sort them in increasing order. One algorithm for doing this, known as *selection sort*, works like this. First, we "cruise" through the list, looking for the smallest element. This takes n steps because the list has length n. At the end of cruising through the list, the algorithm knows the smallest element and it puts that element in the first position in the list by simply swapping the first element in the list with the smallest element in the list. (Of course, it might be that the first element in the list *is* the smallest element in the list, in which case that swap effectively does nothing. In any case, after the swap we are guaranteed that the smallest element is at the beginning of the list.)

In the next phase of the algorithm, we're looking for the second smallest element in the list. In other words, we're looking for the smallest element in the list excluding the element in the first position, which is now known to be smallest.

Thus, we can start cruising from the second element in the list and this second phase will take n − 1 steps. The phase after this will take n − 2 steps, then n − 3, and all the way down to 1. So, the total number of steps taken by this algorithms is n + (n − 1) + (n − 2) + . . . + 1. There are n terms in that sum, each of which is at most n. Therefore, the total sum is certainly less than n × n = n^2. It turns out, and it's not too hard to show, the sum is actually n(n + 1)/2 which is approximately ½ n^2.

A computer scientist would say that the running time of this algorithm grows proportionally to n^2 or is *order* n^2. It's not that the running time is necessarily exactly n^2, but whether it's ½ n^2, 42n^2, or n^2 + n, the n^2 term dominates the running time of the algorithm.

A bit of analysis shows that the memoized algorithm for sequence alignment that we saw in Part II has running time n^2 where n is the length of the longer of the two input sequences.

The UPGMA algorithm that we saw in Part III is also order n^2 where n is the number of given species. And the memoized RNA folding algorithm in Chapter 12 of this part has running time of order n^3. Algorithms that run in time order n^k for some number k are said to run in *polynomial time* because n^k is a polynomial of degree k. Polynomial time algorithms are considered "fast."

14.2 Slow Algorithms

Finding an optimal tour for the Traveling Salesperson Problem is considerably more difficult than the other problems we've explored so far. We've seen that a greedy approach doesn't necessarily give an optimal solution. We could simply enumerate every one of the possible different tours, evaluate the cost of each one, and then find the one of least cost. If we were to take this approach in a situation with n cities, how many different tours would we have to explore?

Notice that there are n − 1 choices for the first city to visit. From there, there are n − 2 choices for the next city, then n − 3 for the third, etc. Therefore there are a total of (n − 1) × (n − 2) × (n − 3) . . . × 1. You may recall from Chapter 5 that the product is called n − 1 *factorial* and denoted "(n − 1)!". In Chapter 5, we wrote a recursive function to compute the factorial and saw just how fast factorials grow. For example, while 5! is a modest 120, 10! is over 3 million and 15! is over 1 trillion. Computers are fast, but examining one trillion different tours would take a long time on even the fastest computer.

Table 14.1. Time to solve problems of various sizes using algorithms with different running times on a computer capable of performing one billion operations per second.

Algorithm running time	Problem size			
	10	20	30	40
n	0.00000001 sec	0.00000002 sec	0.00000003 sec	0.00000004 sec
n^2	0.00000010 sec	0.00000040 sec	0.00000090 sec	0.00000160 sec
n^3	0.00000100 sec	0.00000800 sec	0.00002700 sec	0.00006400 sec
n^5	0.00010000 sec	0.00320000 sec	0.02430000 sec	0.10240000 sec
$n!$	0.00360000 sec	77.1 years	8.4×10^{15} years	2.5×10^{31} years

Table 14.1 demonstrates just how bad a function like n! is in comparison to polynomial functions like n, n^2, n^3, or even n^5. In this table, we are imagining a very fast computer that is capable of performing 1 billion (10^9) "operations" per second. Consider our algorithm for finding the minimum element in a list in time proportional to n. It can find the minimum in a list of 10 elements in $10/10^9 = 10^{-8}$ seconds, and the minimum in a list of 40 elements in $40/10^9 = 4 \times 10^{-8}$ seconds. Similarly, our sorting algorithm that runs in n^2 time can sort 10 elements in $10^2/10^9 = 10^{-7}$ seconds and a list of 40 elements in $40^2/10^9 = 1.6 \times 10^{-7}$ seconds. However, while an algorithm that runs in n! time can solve a problem of size 10 in $10!/10^9 = 0.0036$ seconds, a problem of size 40 takes $40!/10^9 = 2.5 \times 10^{31}$ years!

One might be tempted to believe that in a few years, when computers get faster, an algorithm that runs in n! time might no longer be a problem. While faster computers make a difference for polynomial time algorithms, functions like n! grow so fast that a computer that is 10 or 100 or 1000 times faster provides us with very little relief. For example, while our n! algorithm takes 8400 trillion years for a problem of size 30 on our current 1 billion operations per second computer, it will take 8.4 trillion years on a computer that is 1000 times faster. That's hardly good news!

The Traveling Salesperson Problem is just one of many problems for which algorithms exist but are much too slow to be useful – regardless of how fast computers become in the future. In fact, the Traveling Salesperson Problem is in a group – or "class" – of problems called the *NP-hard problems*. Nobody knows how to solve any of these problems efficiently (that is, in polynomial time like n or n^2 or n^k for any constant k). Moreover, these problems have the property that if any one of them could be solved efficiently (in polynomial time) then, amazingly, all of them could be.

Metaphorically, we can think of the NP-hard problems as a long circle of dominoes – one domino for the Traveling Salesperson Person and one for every NP-hard problem. We don't know how to knock any of these dominoes over (metaphorically, how to solve any of them efficiently) but we know that if any one of them falls over then all of the rest of them will fall as well.

Unfortunately, many important computational problems are NP-hard. For example, while UPGMA was fast, two more biologically "robust" methods for inferring phylogenetic trees called *maximum parsimony* and *maximum likelihood* are both NP-hard. The very important problem of determining how proteins fold into three-dimensional structures is also NP-hard. There are literally thousands of computational problems that are known to be NP-hard.

One approach to dealing with NP-hard problems is to use *heuristic algorithms* (also known simply as *heuristics*). A heuristic is a fast method for solving a problem reasonably well but not necessarily perfectly. For example, there are heuristics for finding maximum parsimony phylogenetic trees that are quite fast but don't guarantee finding the "best" tree. Similarly, there are heuristics for protein folding that find solutions that are good but not necessarily optimal. In the next section we describe a heuristic method called genetic algorithms. The programming problem for this chapter asks you to write a program to solve the Traveling Salesperson Problem using a genetic algorithm.

14.3 Genetic Algorithms

A genetic algorithm is a heuristic technique for dealing with NP-hard problems that takes its inspiration from evolution with natural selection. The idea is to "evolve" reasonably good solutions to a problem by simulating evolution on a population of possible solutions. Returning to our Traveling Salesperson example, let's call the cities in Figure 14.1 by their first letters: A, B, C, D, and E. This is shown in Figure 14.2.

We can represent a tour by a sequence of those letters in some order, beginning with A and with each letter appearing exactly once. For example, the tour Aville to

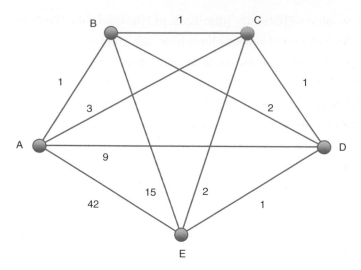

Figure 14.2. The cities in Figure 14.1, using the first letter of each city as its name.

Beesburg to Deesdale to Eetown to Ceefield and back to Aville would be represented as the sequence ABDEC. Notice that we don't include the A at the end
because it is implied that we will return to A at the end.

Now, let's imagine a collection of some number of orderings such as ABDEC,
ADBCE, AECDB, and AEBDC. Let's think of each such ordering as an "organism"
and the collection of these orderings as a "population." Pursuing this biological
metaphor further, we can evaluate the "fitness" of each organism/ordering by
simply computing the cost of flying between the cities in that given order.

Evolution with natural selection depends on mutational processes to produce the
diversity from which it can select. Our genetic algorithm should also include such
processes. Mutating a tour such as ACBDE might involve randomly changing the
order of two adjacent cities. For example, switching the order of D and E in ACBDE
would result in the tour ACBED.

Our simulation starts with a population of organisms/orderings. We evaluate the
fitness of each one. Now, following the biological metaphor, some fraction of the
most fit orderings are allowed to reproduce, resulting in new "child" orderings
where each child has some attributes from each of its "parents" (and potentially
some new attributes introduced by mutation). We now construct a new population
of such children for the next generation. Hopefully, the next generation will be
more fit – that is, it will, on average, have less expensive tours. We repeat this

process for some number of generations, keeping track of the most fit organism (least cost tour) that we have found and report this tour at the end.

What's all this about mating Traveling Salesperson orderings? There are many possible ways we could define the process by which two parent orderings give rise to a child ordering. For the sake of example, we'll describe a very simple method. On the programming problem, you'll be encouraged to try some variants on this method.

Imagine that we select two parent orderings from our current population to reproduce (we assume that any two orderings can mate): ABDEC and ACDEB. We choose some point at which to split the first parent's sequence in two, for example, as ABD|EC. The offspring ordering receives ABD from this parent. The remaining two cities to visit are E and C. In order to get some of the second parent's "genome" in this offspring, we put E and C in the order in which they appear in the second parent. In our example, the second parent is ACDEB and C appears before E, so the offspring is ABDCE. You can think of this as a form of genetic recombination.

Let's do one more example. We could have also chosen ACDEB as the parent to split, and split it at AC|DEB, for example. Now we take the AC from this parent. In the other parent, ABDEC, the remaining cities DEB appear in the order BDE, so the offspring would be ACBDE.

In summary, many important computational problems have no known efficient algorithms. These so-called NP-hard problems are therefore often solved using heuristics which are reasonably fast and find good, but not necessarily the best possible, solutions. Computational biology is replete with NP-hard problems including the prediction of how proteins fold, the simultaneous alignment of multiple sequences (as opposed to aligning just two sequences as we did in Part II), and many others.

Genetic algorithms are one class of heuristics for NP-hard problems. A genetic algorithm is a computational technique that mimics evolution with natural selection. The technique allows us to find good solutions to hard computational problems by imagining candidate solutions to be metaphorical organisms and collections of such organisms to be populations. The population will generally not include every possible "organism" because there are far too many. Instead, the population comprises a relatively small sample of organisms and this population evolves over time until we (hopefully!) obtain very fit organisms (that is, very good solutions) for our problem.

In the programming problem for this chapter, you'll write a program that implements this genetic algorithm for the Traveling Salesperson Problem using both crossover (mating) and mutation. But, rather than using birds or bees to traverse the solutions, you'll use our loyal friend the turtle to walk and draw these tours!

FOR FURTHER READING

Kamil AC and Jones JE. The seed-storing corvid Clark's nutcrackers learn geometric relationships around landmarks. *Nature* 1997;390:276–279.

Lihoreau M, Chittka L and Raine NE. Travel optimization by foraging bumblebees through readjustments of traplines after discovery of new feeding locations. *The American Naturalist* 2010;176(6): 744–757.

Schwartz R. Regulatory network inference. In Pevzner P and Shamir R, eds. *Bioinformatics for Biologists*. Cambridge University Press, 2011.

Where to go from here

In the four parts of this book we've introduced foundational concepts from computer science in the context of biology. We had a good time writing it, and hope you've enjoyed using it.

Over the course of the book you've learned some powerful and fundamental techniques that are used throughout computational biology. You've also learned a valuable general skill – how to design computational solutions and implement them in your own programs. Like other skills, computational problem-solving and programming benefit from practice. As you do more, you'll get even better at it.

With that in mind, we hope that you come away from this book with the confidence to take on new problems. These might range from writing a short program to do some quick analysis, to interfacing with existing programs, to solving altogether new research problems. The key is to find ways to use these tools to further your own interests. In the process, perhaps you'll join us in the excitement that arises when computational techniques are used to explore the many mysteries of life on Earth.

INDEX